CALIFORNIA DESERT WILDFLOWERS

California Desert Wildflowers

by Philip A. Munz

UNIVERSITY OF CALIFORNIA PRESS
Berkeley, Los Angeles, London

UNIVERSITY OF CALIFORNIA PRESS
Berkeley and Los Angeles, California

UNIVERSITY OF CALIFORNIA PRESS, LTD.
London, England

Third Printing, 1972
ISBN: 0-520-00898-7, cloth
0-520-00899-5, paper
Library of Congress Catalog Number 62-8626
Printed in the United States of America

CONTENTS

INTRODUCTION

The Rancho Santa Ana Botanic Garden at Claremont, California, was established for the study of the native plants of California. When therefore in 1959, after about twelve years of continuous work, the large technical book *A California Flora* (by Munz and Keck, University of California Press) was published, it seemed to me that the Botanic Garden as an institution and I as an individual owed something to the layman, the person who has no particular botanical training but who likes to know something about his surroundings in nature. I therefore planned a series of three small books which might be placed in the glove box of the car or carried easily when on a hike. These books were to consist primarily of pictures, some as ink drawings and some as color photographs, with just enough text to give names and a few pertinent facts describing the plants and their location. The young man who made most of the drawings for the first of these three books suggested the catching title "Posies for Peasants," and caught exactly the idea of a nontechnical approach I have tried to imagine.

The first of these books has now been out for a few months. Called *California Spring Wildflowers,* it portrays plants found between the Sierra Nevada and the more southern mountains on the one hand and the sea on the other. It has met such a warm response that I am heartened to present herewith the second one, *California Desert Wildflowers,* for the area east of the Sierra Nevada from Mono County south to northern Lower California. The third one contemplated is *California Mountain Wildflowers* for the pine belt of our higher mountains.

The California Deserts

The California deserts comprise a considerable area if we understand by them the region below the Yellow Pine, beginning in the north with the lower slopes of the Sierra Nevada and with a large part of the Inyo-White Range and its environs, and ending in the south with the Imperial Valley and the arid mountains to the west and the sandy region toward the Colorado River. Roughly and for practical purposes we can think of our desert as consisting of: (1) the more northern Mojave Desert reaching as far south as the Little San Bernardino and Eagle mountains and the ranges to the east, and (2) the more southern Colorado Desert. Being quite different from each other, these two deserts are worth short separate discussions.

In the first place the Mojave Desert, except for the Death Valley region and that about Needles, lies mostly above 2000 feet. Hence, it has more rainfall and colder winters. It opens out largely toward the northeast and in many ways is an arm of the Great Basin of Utah and Nevada and its plant affinities often lie in that direction. The Colorado Desert, on the other hand, consists largely of the Salton Basin, much of it near or below sea level. It opens toward the southeast and its affinity floristically is with Sonora and it is often placed as part of the Sonoran Desert. It is not surprising, then, that many species of the Mojave Desert extend into Nevada and southwestern Utah, while many of the Colorado Desert range into Sonora and western Texas. There are of course many patterns of more limited distribution, such as along the mountains bordering the western edge of the Colorado Desert from Palm Springs into northern Lower California, or around the western edge of the Mojave Desert from the base of the San Bernardino Mounntains to the Tehachapi region.

The climatic conditions on the desert and the situation for plant growth are severe. Plants have had to resort to interesting devices to exist at all. In the first place, seeds of many of them have so-called inhibitors that prevent germination unless thoroughly leached out by more than a passing shower. This means that for many of them it takes a good soaking rain to get started, one that will wet the ground sufficiently for the seedling to send a root down below the very surface. A second characteristic of many of the annuals is that if the season is rather dry, they can form a few flowers even in a most depauperate condition and ripen a few seeds under quite trying circumstances. Thirdly, many of those plants that do live over from year to year, cut down evaporation by compactness—small fleshy leaves, reduced surface as in cacti—by coverings of hair or whitish materials that may reflect light and hence avoid heat, and by resinous or mucilaginous sap that will not give up its water content easily, as exemplified by Creosote Bush and cacti.

A widespread popular fallacy should be mentioned. We read of the great depth to which desert plants can send their roots in order to tap deep underground sources of moisture. This situation is true along washes and watercourses and basins, where Mesquite and Palo Verde, for examples, send roots down immense distances, but on the open desert an annual rainfall of six or eight inches distributed over some months may moisten only the upper layers of soil. Therefore, shrubs like Creosote Bush and plants like cacti tend to have very superficial wide-spreading roots that can gather in what moisture becomes available.

Something should be said too about summer rains. On the coastal slopes at elevations below the pine belt, we are accustomed to summer months practically without rain. But in Arizona and the region to the east of us there are two definite rainy seasons, the one producing a spring flora and the other a late summer and early autumn crop. For the most part the annuals that come into bloom in these two distinct seasons are quite different. Many summers the Arizona rains reach into the desert areas of California and sometimes produce veritable cloudbursts of water. At such times thunderheads appear over the adjacent mountains like the San Bernardino, San Jacinto, and San Gabriel ranges, and the neighboring coastal valleys are much more humid and uncomfortable than when the desert is dry. After these summer rains some of the perennials may exhibit new growth and flowering, and a new crop of annuals may appear, such as Chinchweed and Kallstroemia. I have seen the desert floor green with the last-named plant for miles in early September in the southern Mojave Desert west of Baker and Cronise Valley.

As any desert habitué knows, plant life there is not uniform, but varies with elevation, drainage, character of soil, and the like. One of the characteristic features is the presence of many undrained basins, known locally as "dry lakes," where water may gather in ususually wet years only to dry up more or less completely after a few weeks. Such a situation through the centuries brings about the accumulation of salts or alkali, making these areas too salty for any plant life, or at the fringes there may be an accumulation of species adapted to salty conditions, such as various members of the Pigweed Family: Saltbush, Shadscale, and Glasswort. These basins are scattered over the Mojave Desert and form a series along the old channel of the Mojave River which flows eventually into Death Valley, the largest of all. A similar situation exists in the area near the Salton Sea.

The great open plains and flats of much of the desert are covered with Creosote Bush (which is associated with Burroweed), Boxthorn, Incienso, and many other species. Here the average rainfall is from two to eight inches and summer temperatures may be very high. Some cacti grow in this region, which mostly is pretty well drained, but many are found on rocky canyon walls, in stony washes, and other places also. In areas above the Creosote Bush on the Mojave Desert, say from 2,500 feet to 4,000 or higher, Joshua Trees tend to distribute themselves in a sort of open woodland with lower shrubs in between. Here the annual precipitation may be from six to fifteen inches and the vegetation is correspondingly richer. And then, along the western edge of the Colorado Desert and more particularly in the mountains bordering on and situated in the Mojave Desert is a zone of Pinyon and Juniper, mostly at 5,000 to 8,000 feet. Here the annual precipitation runs about twelve to twenty inches a year, some of it as snow. This belt has some summer showers and some plants in bloom in the summer and even into fall, as well as in the spring, which comes later than in the Creosote Bush zone. Particularly in the more northern parts of the desert, Creosote Bush gives way in the upper elevations to Sagebrush (*Artemisia tridentata* and relatives), and large regions in Lassen, Mono, and northern Inyo counties have a Sagebrush desert like that of Nevada and Idaho. With so wide a diversity of conditions, then, it is not surprising to find quite different flowers at various altitudes and in various habitats.

How to Identify a Wildflower

To refresh the reader's memory, a drawing is presented (figure A) showing the parts of a typical flower, since in the text it is impossible to talk about plants and their flowers without using the names of some of the parts. In the typical flower we begin at the outside with the *sepals*, which are usually green although they may be colored. The *sepals* together constitute the *calyx*. Next comes the *corolla* made up of separate *petals*, or the petals may fuse forming a tubular or bell-shaped corolla. Usually it is the conspicuous part of the flower, but it may be reduced or lacking altogether, and its function of attraction of insects for pollination may be assumed by the *calyx*. Then as we proceed inward in the flower, we usually find the *stamens*, each consisting of an elongate *filament* and a terminal *anther* in which the pollen is formed. At the center are one or more *pistils*, each with a basal *ovary* containing the ovules or immature seeds, a more or less elongate *style*, and a terminal *stigma* with a rough sticky surface for catching pollen. In some species, stamens and pistil are borne in sep-

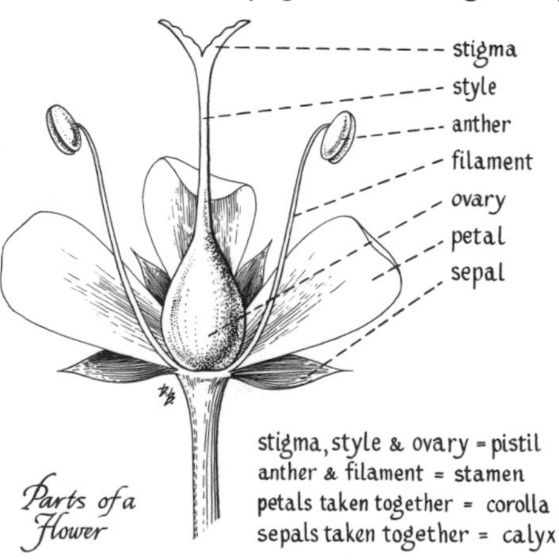

Figure A. A Representative Flower

arate flowers or even on separate plants. In the long evolutionary process by which plants have developed into the many thousands of types of the present day and have adapted themselves to various pollinators, their flowers have undergone very great modifications and so now we find more variation in them than in any other plant part. Classification is thus largely dependent on the flower.

To help the reader identify a flower, the drawings are grouped by color, but in attempting such an arrangement it is difficult often to place a given species where everyone would agree it belongs. Color varies so much from deep red into purple, for example, and from blue into lavender, from whitish into greenish, from yellow through cream toward white, that it is impossible to satisfy one's self, let alone his readers. I have done the best I could, after consultation with others who know the plants, and have tried for the general impression given, especially when flowers may be very minute and the color effect may be caused by other parts. My hope is, however, that by comparing a given wildflower with the picture it seems to resemble and then by checking against the few facts given in the text, the user of this little book may be able in most cases to come to some conclusion as to what his particular plant may be.

One of my chief difficulties in writing such a book is to find usable common names. I am not interested in taking those coined from the scientific name by a professional botanist who breaks down the genus name into its Greek roots and then adds the species name, such as for example, the common name "Mrs. Ferris's Club Flower" for *Cordylanthus Ferrisianus.* On the other hand, if the local inhabitants call this Bird's Beak, that is acceptable. But for some plants of interest which my readers may find and for which they wish to know a name, I may have been unable to ascertain a true folk-name. In these cases it has seemed best to me to use the genus name, like Phacelia or Oxytheca, as a common name too. Many desert plants are not very conspicuous and just seem not to have good widely used names, so we have to resort to this scientific appellation.

Another problem confronting the writer of such a book is to select what plants to present in it. I have tried for the most part not to include those already shown in *California Spring Wildflowers,* so that anyone having both books will have that many more species shown. I have chosen plants in which I feel there may be an interest, and that does not necessarily mean that all should be showy or common. I have attempted not to present the various closely related forms of a complex group, but have taken one as an example and often added a word about additional forms closely resembling the first. I have tried too to include species from the different parts of the desert, so that the book will be useful for more than just the Palm Springs region, for example, or just for Death Valley, to name two of the commonly visited areas.

I cannot help but register a plea that residents of the desert and visitors thereto exercise discretion in picking, transplanting, and otherwise interfering with normal development and reproduction of desert plants. The thousands of people who live in or visit the desert nowadays are bound to inflict hardship on the vegetative covering. With the scant rainfall, desert plants grow slowly and a branch broken off a pinyon tree for a campfire may have taken many years to produce. Certainly those who have known the desert over a period of years cannot help but be appalled by the magnitude of the recent destruction.

The scientific nomenclature used in this volume is that employed in the more technical work by Munz and Keck, *A California Flora,* University of California Press, 1959. Reference to that book is recommended when more information is sought than is here available.

ACKNOWLEDGMENTS

Most of the drawings used in this book were made by various graduate students working at the Rancho Santa Ana Botanic Garden: Dick Beasley, Stephen Tillett, and Shue-Huei Liao. Others were by Helen G. Laudermilk. Still others by Milford Zornes, Rod Cross, and Tom Craig were used in 1935 in my *Manual of Southern California Botany* which has been out of print for many years. This book was copyrighted by Claremont College and I wish to thank President Robert J. Bernard of that institution for permission to reproduce these drawings now. The Kodachromes belong to the collection of the Rancho Santa Ana Botanic Garden, many of them having been taken by Percy C. Everett. It is a pleasure to acknowledge the help of all those mentioned above and of Gladys Boggess in preparation of manuscript.

Philip A. Munz
Rancho Santa Ana Botanic Garden
Claremont, California
June 1, 1961

FERNS AND CONE-BEARERS

Section One

Ferns, as everyone knows, do not reproduce by flowers and many-celled seeds which are borne in seed-pods, but usually have on the under surface of the leaves roundish or elongate "fruit-dots" or sori. From these are released the one-celled microscopic spores instead of seeds. Ferns are usually associated with fairly moist situations and are hardly expected on the desert. If there, they are most apt to be under overhanging rocks and often in rock-crevices where they are protected from browsing and where they are shaded and get full benefit of the scanty rainfall. One of the most common desert ferns is CLOAK FERN (*Cheilanthes Parryi*), figure 1, a low tufted plant. The leaves have a light gray or brownish wool on their under surfaces and are borne on wiry purplish-brown petioles or stipes. Cloak Fern is found at elevations below 7,000 feet, from the White Mountains to the Colorado Desert.

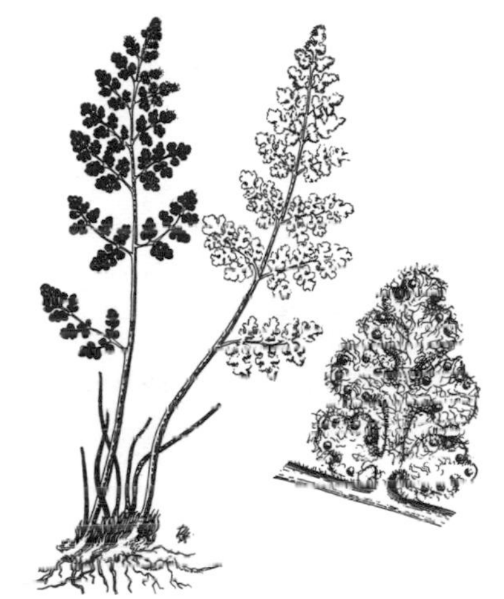

FIGURE 1. CLOAK FERN

Common on the California deserts in rocky places below 9,000 feet is another plant of about the same size and habit, but with the fronds covered with overlapping scales beneath instead of wool. It is LIP FERN or BEAD FERN (*Cheilanthes Covillei*), figure 2, the individual segments of the frond being so inrolled at the margins that from above they look like little green beads.

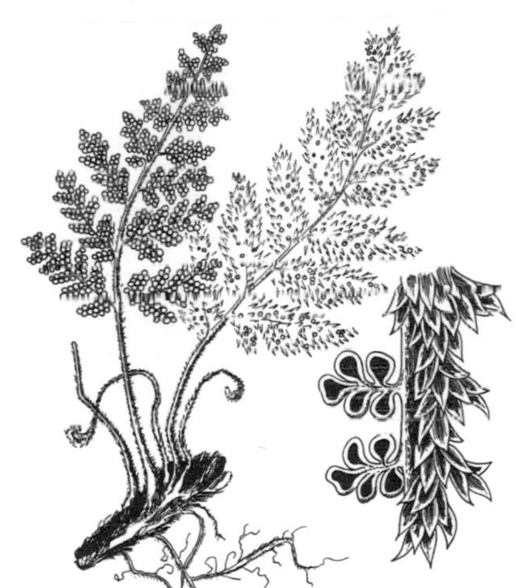

FIGURE 2. BEAD FERN

The DESERT GOLDENBACK (*Pityrogramma triangularis* var. *Maxoni*), figure 3, constitutes a third fern with low tufts. Its frond is wider in proportion to length than are those of the other two, and is yellowish- or whitish-powdery beneath instead of woolly or scaly. All three of these are apt to be found in rocky canyons like those near Palm Springs. During the dry season they roll up tightly and form dusty little clumps.

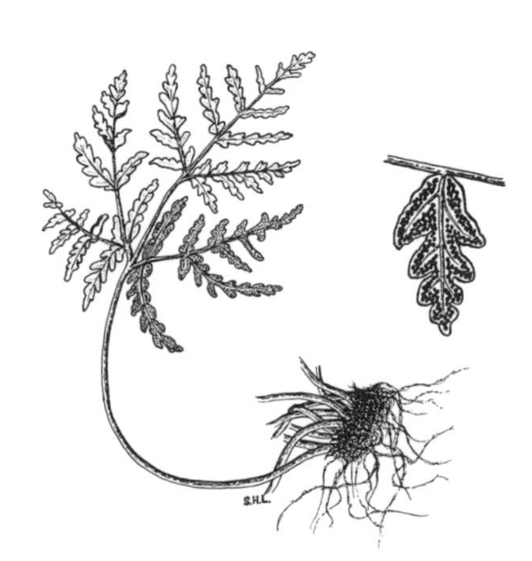

FIGURE 3. GOLDENBACK

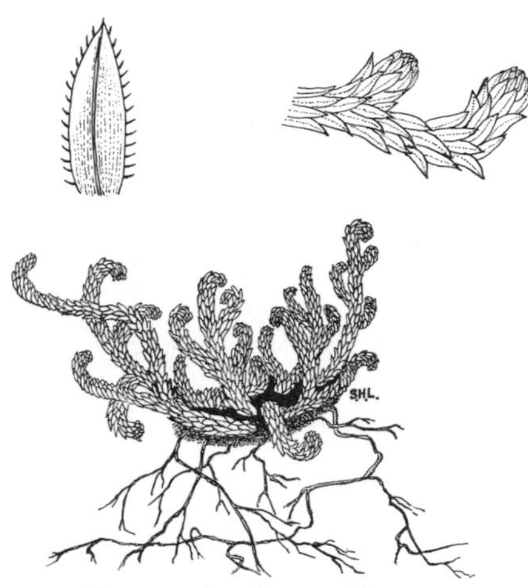

Figure 4. Spike-Moss

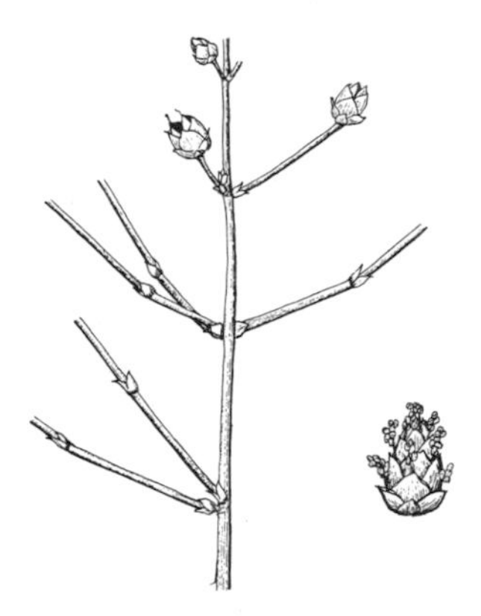

Figure 5. Mormon-Tea

Often associated with the ferns in not producing flowers and seeds, but multiplying by single-celled spores, are the Little Club-Mosses or Spike-Mosses. These plants belong to the genus *Selaginella.* Our California species are low or creeping, with minute scalelike overlapping leaves. The tips of some branches have the leaves slightly modified and bearing solitary sporecases in their axils. Often we can see these only by bending the leaves away from the stem. On the desert we have two most characteristic species of Little Club-Moss, the common one below 3,000 feet, in canyons along the western edge of the Colorado Desert, in the Chuckawalla Mountains, and other similar places. It is *Selaginella eremophila,* figure 4, with the stems quite flat on the ground. The other, S. *leucobryoides,* is cushionlike, in rocky places between 2,000 and 7,500 feet, in such ranges as the Panamint and Providence mountains. It is distinguished also by having a bristle at the tip of each leaf.

The next group of plants after the ferns and their allies is that of the cone-bearing trees, including the pines and junipers as examples. Common on higher slopes in the desert are the Pinyon and the Juniper or Cedar. But over most of the desert and distantly related to the foregoing is a shrubby group, Mormon-Tea or Mexican-Tea, of which the species *Ephedra nevadensis,* figure 5, is illustrated. The stems are jointed, with scalelike leaves in two's or three's (according to which species we have), some plants bearing seed-producing cones, as attached to the twigs in the illustration, and others bearing male cones with clusters of stamens projecting from above the cone-scales.

FLOWERS ROSE TO PURPLISH-RED OR BROWN

Section Two

Figure 6. Wild Onion

Wild Onion is the name of a large group of bulbous plants with an onion or garlic odor. One of the most common species on the desert is *Allium fimbriatum*, figure 6, with a rounded bulb of which the outer coats are dark. The leaf is solitary and rounded in cross section; the flower cluster is compact, the flowers themselves rose to purple with darker midveins. There are many forms of this species, varying in height, but mostly extending a very few inches above the ground. Found on dry slopes and flats between 2,000 and 8,000 feet, this Onion is in the deserts and on their western edge, reaching the Coast Ranges.

In the Buckwheat Family (see pages 30, 91), we have Chorizanthe, the species shown here, *Chorizanthe Thurberi*, figure 7, being erect, forked above, and with basal leaves. The flowers are very small and contained in an involucre with three basal spreading horns. These involucres are often quite highly colored. They end at the summit in spine-tipped teeth. The species is common in dry sandy places below 7,000 feet throughout our deserts and east to Utah and Arizona. To the west it ranges north to San Benito County. Flowering is from April to June.

Figure 7. Chorizanthe

Amaranth is related to the Garden Cockscomb, the common desert species being Fringed Amaranth (*Amaranthus fimbriatus*), figure 8, a handsome annual with slender erect stems to two feet high and with numerous narrow leaves. The small flowers are many, without petals, but the rose or lavender sepals are fimbriate on the edge and make conspicuous masses of color. Found in dry gravelly places below 5,000 feet, the species occurs from the Colorado and eastern Mojave deserts to Utah and Mexico. It is an autumn plant.

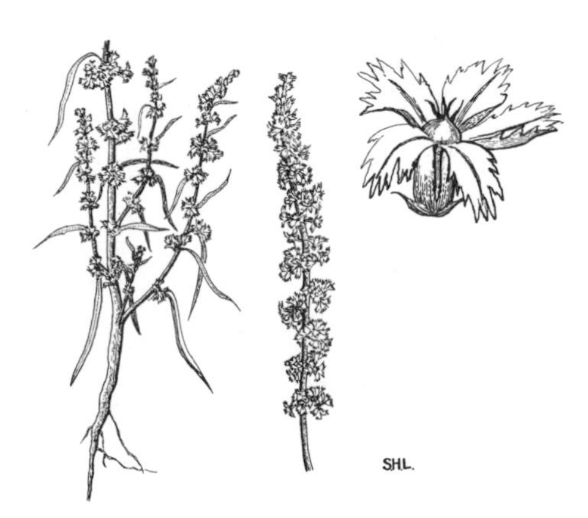

Figure 8. Fringed Amaranth

Figure 9. Sand-Verbena

Figure 10. Sticky Ring

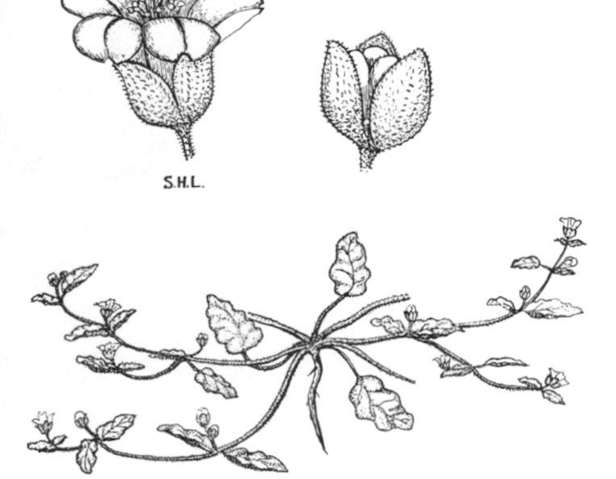

Figure 11. Windmills

Sand-Verbena is not related to the true Verbena at all and its resemblance is quite superficial. It belongs to the Four-O'Clock Family (see page 31). Sand-Verbena is of several kinds, one of the more common being *Abronia villosa,* figure 9, a much-branched annual with stout stems to almost two feet long, hairy, often viscid. The clusters of flowers are subtended by narrow bracts. The purplish-rose flowers are about half an inch long and very fragrant. The fruit is hard and winged so as often to be more or less notched on top. The plant is common in open sandy deserts and interior coastal valleys and eastward to Nevada and Sonora. It flowers from February to July.

In the same family and from the Death Valley region is found a remarkable coarse perennial herb with one to few stems one to three feet high. They are marked with transverse reddish sticky bands and support above an open inflorescence with slender branches which end in headlike clusters of small pinkish to greenish flowers. Sometimes called Sticky Ring, sometimes Boerhaavia, the plant is *Boerhaavia annulata,* figure 10, of which the flower and fruit are shown in the lower left-hand corner. The species is met in sandy and gravelly places below 3,000 feet and blooms in April and May.

Another member of the same family is the trailing plant Windmills or Trailing Four-O'Clock (*Allionia incarnata*), figure 11. It is mostly a perennial with viscid stems along which are scattered the flowers in clusters of three. They are rose-magenta and about one-half inch or more in diameter. Found on dry stony benches and slopes below 5,000 feet, Windmills is a characteristic plant of both deserts, ranging east to Colorado and Chihuahua. Flowers from April to September.

In the same family with the Ice Plant the desert has a native plant called SEA-PURSLANE (*Sesuvium verrucosum*), figure 12. It is a rather fleshy freely branched perennial with stems almost flat on the ground. The leaves are in pairs (opposite), spatulate, fleshy, and with the flowers in their axils. The sepals are purplish to rose-pink, with transparent margins. It is not a beautiful plant, but is quite common, especially in more or less saline places. Flowers appear from April to November.

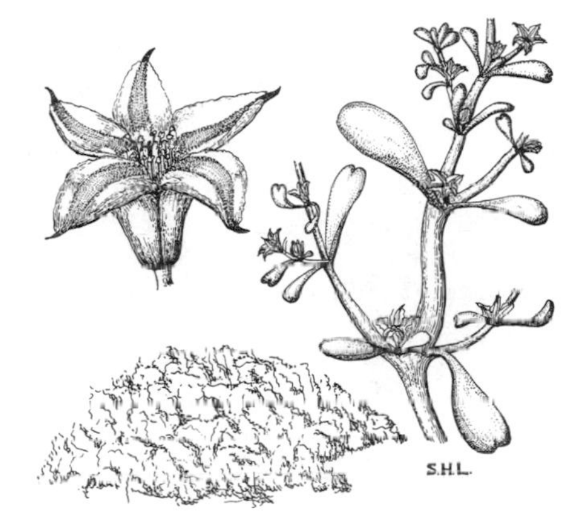

FIGURE 12. SEA-PURSLANE

In the Buttercup Family (see page 83) the desert surprises us with a lovely little pink ANEMONE (*Anemone tuberosa*), figure 13. The underground part is tuberous; the stem or stems four to twelve inches high; the leaves few, divided into three parts. The rose-colored flower is about an inch across. An inhabitant of dry rocky slopes between 3,000 and 5,000 feet, this Anemone is found in the southwestern Colorado Desert and the eastern and northern parts of the Mojave to Utah and New Mexico. It blooms in April and May.

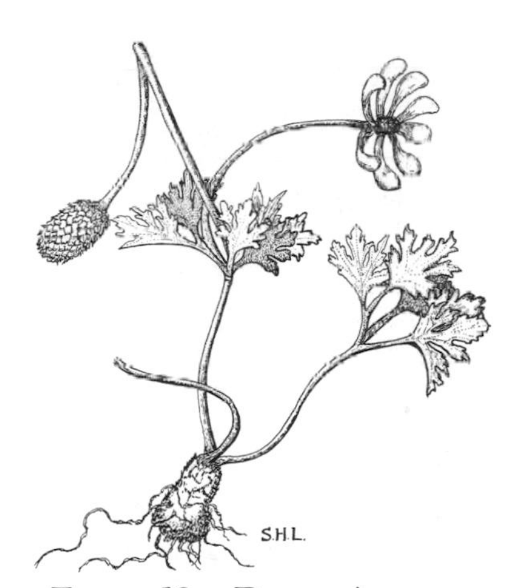

FIGURE 13. DESERT ANEMONE

Another desert plant that is to me altogether fanciful is DESERT CANDLE or SQUAW-CABBAGE (*Caulanthus inflatus*), figure 14, of the Mustard Family (see pages 33, 67, 93). Erect, usually unbranched, it is an annual with curiously inflated hollow stems commonly one to two feet high, bearing numerous leaves below and many flowers above. The sepals are purplish or white with purplish tips, but the cluster of purplish buds at the summit is particularly noticeable. The petals themselves are white. Common on open flats and among brush below 5,000 feet, from the Barstow region west and into the San Joaquin Valley as far as Fresno County. Flowers March to May.

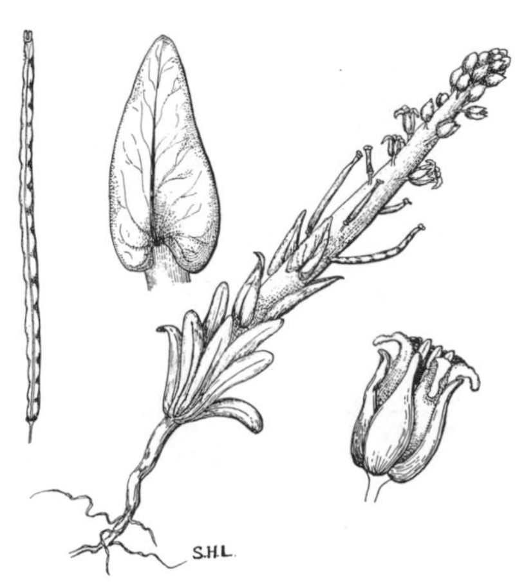

FIGURE 14. DESERT CANDLE

FIGURE 15. ROCK-CRESS

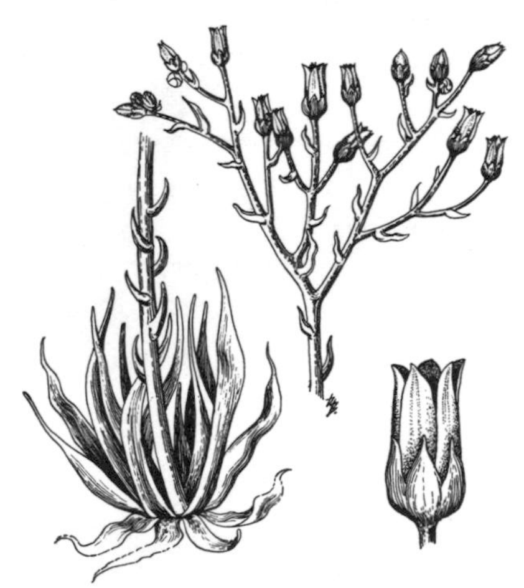

FIGURE 16. LIVE-FOREVER

FIGURE 17. ARIZONA LUPINE

Another member of the Mustard Family is a broad-podded ROCK-CRESS (*Arabis glaucovalvula*), figure 15, a perennial from a somewhat woody base, more or less hoary throughout and six to fifteen inches high. The pubescent sepals are about one-sixth of an inch long, the rather pink or purplish petals slightly longer. The unusual pods are flat, an inch or more long and about one-fourth inch wide. The species is found in dry stony places between 2,500 and 5,300 feet, from Bishop Creek, Inyo County, to the Eagle Mountains, Riverside County, blooming from March to May. More common, perhaps, are other species of Rock-Cress, some with purple, some with reddish flowers, and with narrower pods.

In this day of the popularity of succulents, it is nice to find a LIVE-FOREVER (*Dudleya saxosa*), figure 16, growing wild on the desert. It occupies dry stony places between 3,000 and 7,000 feet in the Panamint Mountains and, in a somewhat modified form, desert slopes in San Bernardino County and south to the Laguna Mountains of eastern San Diego County. The petals are actually yellowish, but the reddish sepals and stems give a general reddish appearance (see page 34). Flowering is from May to June.

The Pea Family (pages 83, 96), with its characteristic pod like that of the cultivated Pea or Bean, is richly represented in California, especially by the lupines. One of these, the ARIZONA LUPINE (*Lupinus arizonicus*), figure 17, is a rather fleshy branched annual. The flowers are pale purplish-pink, often drying violet, and are almost half an inch long. It is common in sandy washes and open places below 2,000 feet, from eastern Inyo County south and into Nevada and Sonora. It blooms commonly from March to May.

One of the largest genera of plants known, *Astragalus,* contains about 2000 species, of which over 400 occur in North America, some under common names such as Locoweed and Rattleweed. It is in the Pea Family and frequent in the desert. The species of LOCOWEED in figure 18 (*Astragalus Layneae*), is a perennial with a deep-seated root, hairy somewhat grayish foliage, and purple-tipped flowers over one-half inch long. The pods are sickle-shaped, one to two inches long. This plant is found in sandy places, often in large colones, between 1,500 and 5,000 feet, over much of the Mojave Desert, from Twentynine Palms to Owens and Death valleys and blooms from March to May. See also page 37.

FIGURE 18. LOCOWEED

KRAMERIA (*Krameria parvifolia* var. *imparata*), figure 19, is an intricately branched thorny little shrub, a foot or so high and of wider spread, and silky-woolly on young growth. The leaves are linear, to about half an inch long. Flowers are red-purple. The pod is short, armed with spines that are barbed in their upper parts. Rather common in rocky and sandy places, Krameria occurs at mostly 2,000 to 4,000 feet, in mountains from the Death Valley region through the eastern Mojave Desert to the Colorado Desert and beyond. A closely related species (*K. Grayi*) has the barbs at the very summit of the spines in umbrella fashion.

FIGURE 19. KRAMERIA

FAGONIA (*F. californica*), figure 20, has a fruit shaped much like that of Krameria, but not spiny. It too is a low intricately branched little plant, each leaf having three leaflets. The flowers are purplish, up to about one-third inch long. Common chiefly on rocky slopes below 2,000 feet, it occurs through most of the Colorado Desert and into the southern Mojave, extending its range also into parts adjacent.

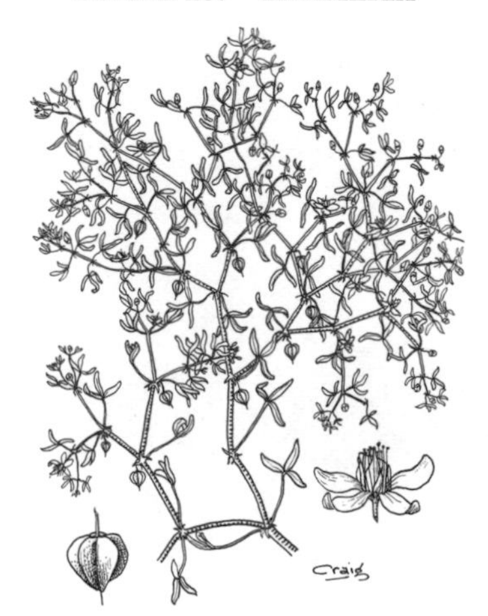

FIGURE 20. FAGONIA

FIGURE 21. HERON BILL

FIGURE 22. TURPENTINE-BROOM

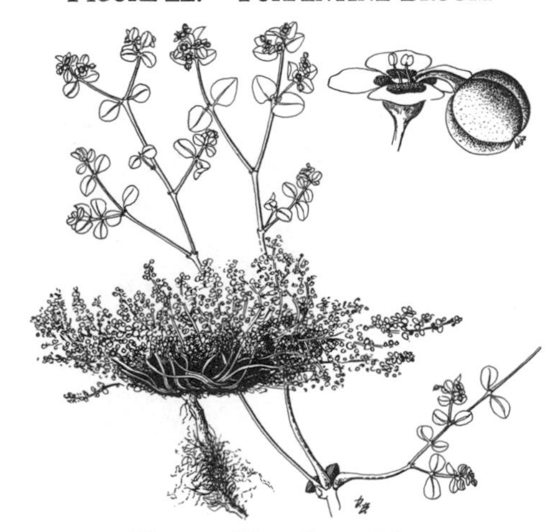

FIGURE 23. SAND MAT

In the Geranium Family is Filaree or Clocks, largely known to us by several species introduced from the Old World. But a native is the desert HERON BILL (*Erodium texanum*), figure 21, an annual with almost prostrate stems and three-lobed leaves. The sepals are silvery with purple veins; the petals purple, about half an inch long. Growing eastward from near Riverside, it is common in dry flats and open places below 3,000 feet, from the eastern Mojave Desert to Lower California and Texas. The flowers come in the spring months.

TURPENTINE-BROOM (*Thamnosma montana*), figure 22, is in the same family with Citrus and like it is covered with oil-bearing glands. It is strong-scented, rather woody, with branching, broomlike, yellowish-green stems one to two feet high. The narrow leaves are shed early. The flowers are purplish, about one-half inch long; the capsule is two-lobed. It is not strange that so aromatic a plant was used by the Indians for medicinal and other purposes, with the idea that it aided healing. Growing on dry slopes below 5,500 feet, it is distributed from Inyo to Imperial and San Diego counties and east to Utah and New Mexico. The flowering season is spring.

For the Spurge Family see pages 70, 71. In figure 23 is shown a GROUND SPURGE or SAND MAT (*Euphorbia polycarpa*), a milky-juiced perennial that occurs in smooth or hairy forms. The flowers have no sepals or petals, but are borne in an involucre on the edges of which are conspicuous white appendages with maroon glands. The species is common on the desert floor below 3,000 feet and reaches into the coastal drainage as far north as Ventura County, going east also to Nevada and Sonora. Flowers can be found over many months; both staminate and pistillate are shown upper right, projecting from the involucre.

TAMARIX (*Tamarix pentandra*), figure 24, is an introduced shrub that has become widely established along water courses and in low, often alkaline places below 4,000 feet. It is a loosely branched shrub or small tree with minute scalelike leaves and whitish or pinkish minute flowers in elongate clusters grouped in great panicles. A number of species have established themselves both in the deserts and coastal areas. Flowers can be found much of the year. The Athel, so commonly planted as a windbreak in desert valleys, is *Tamarix aphylla*.

FIGURE 24. TAMARIX

DESERT-MALLOW OR APRICOT-MALLOW (*Sphaeralcea ambigua*), figure 25, is a perennial with thick woody crown and woody lower stems that grow to be one to three feet tall. It is covered with scurfy grayish or yellowish pubescence. The flowers are grenadine to peach-red, up to one and one-half inches long. It belongs to the Hollyhock Family and is one of our most characteristic desert plants, being widely distributed in a variety of forms and colors and growing below 4,000 feet. It blooms from March to June.

FIGURE 25. DESERT-MALLOW

Closely related to Desert-Mallow is DESERT FIVESPOT (*Malvastrum rotundifolium*), figure 26, an erect annual up to almost two feet tall, simple or branched, and stiff-hairy. The corolla tends not to open up very far, but to remain globular, and one has to look inside to see the five dark spots within the rose-pink to lilac flower. As it is frequent in washes and on mesas below 4,000 feet, this charming plant extends its range over most of our desert areas and into Arizona and Nevada. It blooms from March to May.

FIGURE 26. DESERT FIVESPOT

OCOTILLO OR CANDLEWOOD (*Fouquieria*

FIGURE 27. OCOTILLO

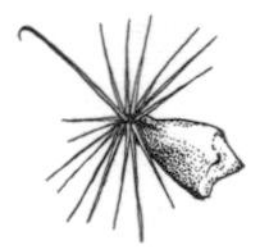

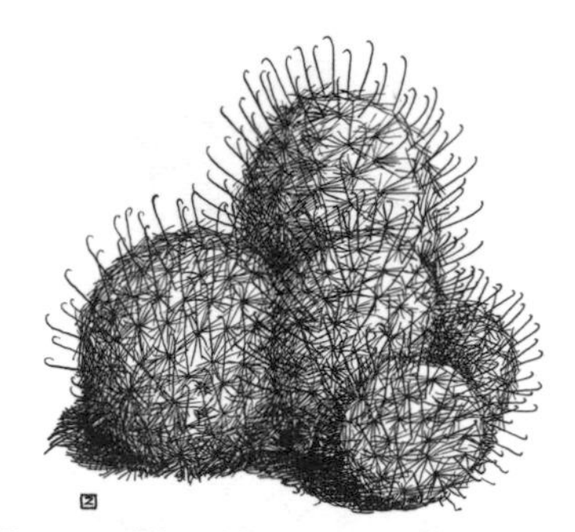

FIGURE 28. FISHHOOK CACTUS

FIGURE 29. DESERT-PARSLEY

splendens), figure 27, is one of the most conspicuous and characteristic shrubs of the desert below 2,500 feet and from the area east of Daggett to Texas and Lower California. Its dry stout stems get to be twenty feet tall and are characteristically furrowed and bear stout divaricate spines that have developed from the petioles of the principal leaves. Secondary leaves appear in fascicles in the axils of these spines after both spring and late summer rains. The scarlet flowers are about an inch long and are very conspicuous in a wet season.

The cacti (see pages 39, 41, 99), grow in the desert in well-drained places and are of quite a variety of forms and sizes and colors. Among these is the cushion type with low, thick, sometimes clustered stems. Such is the FISHHOOK or CORKSEED CACTUS (*Mammillaria tetrancistra*), figure 28. It is mostly a few inches high, with one or more of the central spines hooked and the petals with rose to lavender stripes. The scarlet fruit gets to be one-half inch or more long and is quite persistent. The species is occasional on dry slopes below 2,000 feet in both our deserts and reaches Utah and Arizona, flowering in April.

In the Carrot Family (see page 100), a characteristic desert plant is LOMATIUM or DESERT-PARSLEY (*Lomatium mohavense*), figure 29, an almost stemless hoary-pubescent perennial with finely divided leaves having crowded segments. The flowers are small, mostly purplish, and the fruits are flattened with a wing on either edge. It is fairly abundant on dry plains at 2,000 to 6,000 feet, along the western edge of the Colorado Desert and in the Mojave northwest to Mt. Pinos, Independence, and Death Valley. It flowers in April and May.

It is a great temptation when crossing the sand dunes between Yuma and Imperial Valley to get out of the car and wander about. The person who does so is very apt to find on the surface of the sand an occasional woolly mass attached to a fleshy elongate underground stem and bearing purplish flowers embedded in the terminal mass of wool. This is SAND FOOD (*Ammobroma sonorae*), figure 30, so called because the Indians used the fleshy parts for food, raw or roasted. It is a parasite on the roots of other plants and is found both here and in Arizona and Sonora. A separate flower in its wool is shown at lower right.

Gentians are not common on the deserts, but a member of the family, CANCHALAGUA (*Centaurium venustum*), figure 31, is met along the western edge of the desert, although it is more common in the valleys and on slopes draining toward the coast. It is a low pink-flowered annual with red spots in a white throat and with unusually twisted anthers. Another member of the family with larger flowers up to almost an inch long is Eustoma (page 43). Its petals are deep lavender to bluish and it is occasional in moist places, since its range extends from coast to coast across the South.

DESERT CALICO (*Langloisia Matthewsii*), figure 32, is a member of a family well represented in the West, the Phlox Family (see pages 21, 44, 84). Low, branched and tufted, it has leaves an inch or so long with bristle-tipped teeth. The corolla is two-lipped, mostly pinkish and with a red and white pattern or markings. It is common in great masses in sandy and gravelly flats below 5,000 feet, in the deserts from Inyo County to Imperial County and to Nevada and Sonora. The flowers appear from April to June, often

FIGURE 30. SAND FOOD

FIGURE 31. CANCHALAGUA

FIGURE 32. DESERT CALICO

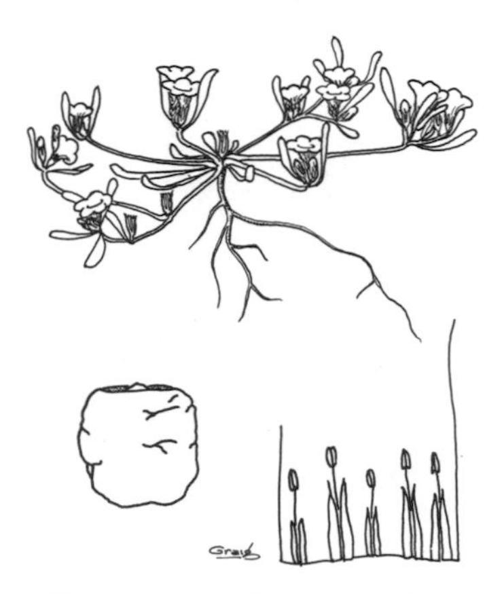

FIGURE 33. PURPLE MAT

FIGURE 34. OWL'S-CLOVER

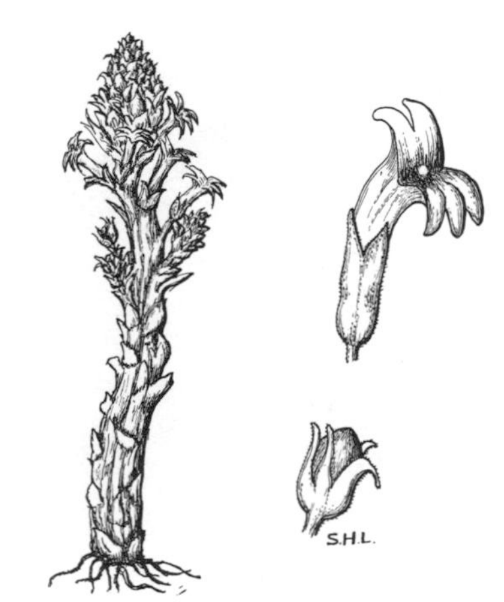

FIGURE 35. BROOM-RAPE

giving a pinkish tinge to areas of some extent. See also page 45.

PURPLE MAT (*Nama demissum*), figure 33, is a good name for another small prostrate annual of dry flats and slopes. In a good year it is a floriferous plant some inches across, in a dryer one it is a tiny tuft. The corolla is purplish-red, almost half an inch long. There are several forms and related species, so that it and its close relatives are found in much of our desert area and in bloom in the spring months. The range is into the deserts to the east of California. The peculiar seed and the scales at the base of the stamens inside the corolla are shown.

In the Figwort Family a characteristic group is the Paint-Brush, an inclusive name used for a number of quite different plants (see page 51). A representative member is PAINT-BRUSH OR OWL'S-CLOVER (*Orthocarpus purpurascens* var. *ornatus*), figure 34, found locally and in profusion in open flats of the western Mojave Desert at between 2,000 and 3,000 feet. Its flowers are more intensely colored than the coastal forms of the species and have a deep red-purple spike, the corolla being deep velvet-red, the outer third of the lower lip orange-yellow. It is a spring bloomer and, with the masses of Gilia and other annuals with which it grows, often presents a brilliant spectacle about Adelanto and into the Antelope Valley.

A plant parasitic on the roots of other plants and also with two-lipped flowers is BROOM-RAPE OR OROBANCHE (*Orobanche ludoviciana* var. *Cooperi*), figure 35. It is quite fleshy, often grows in clusters, and may attain a foot in height. The plant is purplish-gray, contains no chlorophyll, and has reduced scalelike leaves. The purplish flowers are more than half an inch long. It is largely parasitic on Burro-weed (page 106), and may be damaging to tomatoes and other crops in cultivated valleys.

In the Bignonia Family, hence related to the eastern tree Catalpa and to various cultivated vines, is DESERT-WILLOW (*Chilopsis linearis*), figure 36, a large shrub common along washes and in watercourses below 5,000 feet. With willowlike leaves, it produces large flowers over an inch long, deep lavender to whitish (or in the Victorville-Adelanto region a more deeply colored form in cultivation) with purplish lines and markings. It has hanging seed-pods which can be almost a foot long. The flowers come between May and September. The shrub is deciduous during the colder parts of the year.

FIGURE 36. DESERT-WILLOW

CHUPAROSA (*Beloperone californica*), figure 37, is our only member of the large tropical Acanthus Family. It is an often almost leafless shrub one to three feet high, with gray-green twigs and scattered dull scarlet tubular flowers more than an inch long. Hummingbirds visit it constantly, therefore its common name, which in Spanish has reference to sucking. Chuparosa is found along sandy watercourses below 2,500 feet, from the northern and western edges of the Colorado Desert to Lower California and Sonora. It blooms from March to June. See also page 51.

FIGURE 37. CHUPAROSA

THREAD STEM (*Nemacladus rubescens*), figure 38, is overlooked by many persons, because its slender stems and small flowers make it exceedingly inconspicuous against its sandy background. Often it is only by a gust of wind that one becomes aware that something on the desert surface is moving; then closer examination reveals tiny plants never more than a few inches high, with a few basal leaves and yellowish flowers with purple-brown

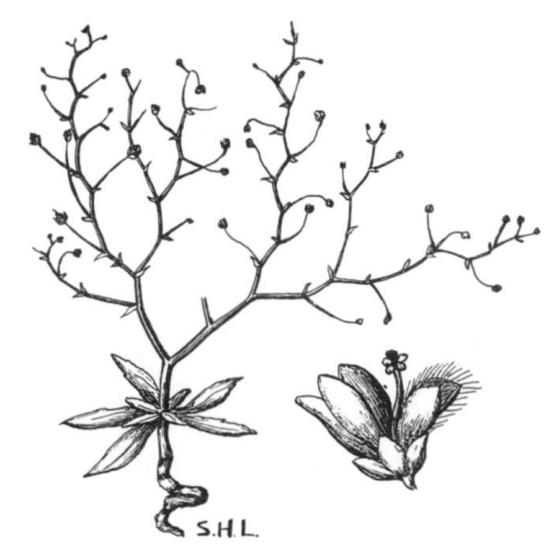

FIGURE 38. THREAD STEM

FIGURE 39. MARSH-FLEABANE

FIGURE 40. SPANISH NEEDLES

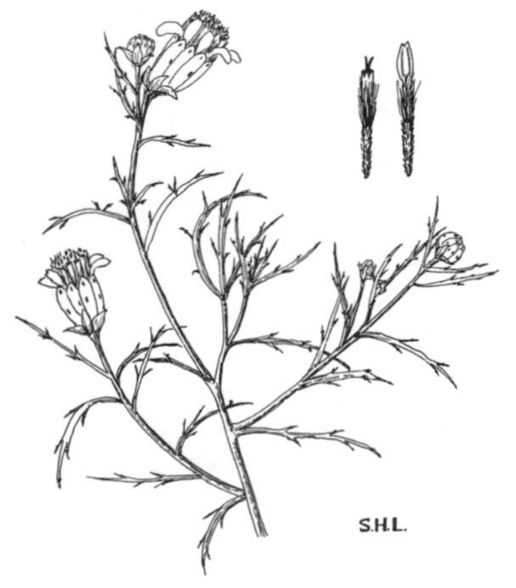

FIGURE 41. HOLE-IN-THE-SAND

markings. Other species may have other colors. Growing below 4,000 feet, this plant is widely distributed over our and adjacent deserts and blooms mostly in April and May.

MARSH-FLEABANE (*Pluchea purpurascens*), figure 39, is an erect herb to four feet high, branched above, green, with leaves two to four inches long. It has large terminal clusters of small heads made up of many small purplish florets. Each seedlike fruit bears a tuft of white hairs. Found in both freshwater and salt marshes, from San Francisco and the Central Valley south, the species occurs also in wet places on the desert from Inyo County south, and to the Atlantic Coast. It blooms from July to November.

SPANISH NEEDLES (*Palafoxia linearis*), figure 40, is a harsh herb, erect, branching, one to two feet high, with somewhat grayish leaves and terminal heads of small pinkish-white tubular florets, of which a single one is drawn separately. Growing on sandy flats and in washes below 2,500 feet, it is a common desert annual, ranging as far east as Utah and northern Mexico and flowering from January to September. On the sand dunes east of Imperial Valley is a larger plant, often perennial and woody at the base.

HOLE-IN-THE-SAND OR NICOLLETIA (*Nicolletia occidentalis*), figure 41, is a low perennial atop a deep-seated taproot and with pinnately-parted rather fleshy strong-scented leaves. The heads of purplish flowers are surrounded by an involucre with evident glands (compare page 108). Found in shallow basins in sand at 2,500 to 4,500 feet, it ranges along the western borders of the Majove and Colorado deserts and flowers from April to June. It is named for an early American explorer, J. N. Nicollet.

COLOR PLATES

Desert-Lily (*Hesperocallis undulata*), plate 1, is one of the most outstanding of the desert wildflowers. The tunicated bulb is deep in the ground and sends up in a damp spring a straight stout simple stem that may become several feet high. Just above the ground is a cluster of elongate leaves of a blue-green color but with white margins. The flower cluster itself may be a foot or more long, each flower, about two inches long, having a silvery-green band on the back of each petallike segment. Desert-Lily is common on dry sandy flats below 2,500 feet, on the Mojave Desert from Yermo east and on the Colorado Desert. It blooms from March to May.

Plate 1 Desert-Lily

Nolina is yuccalike and obviously related to the yuccas in being woody cespitose below and having numerous thick elongate leaves. But its flowers are much smaller and they persist long after anthesis as dry papery conspicuous remains. The common species, *Nolina Parryi*, plate 2, has the leaves saw-edged, while the other desert species, *N. Bigelovii*, has smooth margins. The flower stalks attain a height of several feet. Nolina usually grows in rocky places from the Little San Bernardino and Kingston mountains to the western edge of the Colorado Desert, and on the coastal slopes from Ventura County south. Flowering is from April to June.

Plate 2. Nolina

Blue Dicks or Desert-Hyacinth (*Brodiaea pulchella* var. *pauciflora*), plate 3, is another bulbous plant. It has basal leaves to a foot or more long and a slender naked flower stalk one to two feet high. The flowers are few in a cluster and pale blue. It grows in dry open places over the California deserts, blooming from March to May. On coastal slopes it is represented by the darker blue typical form of the species with more flowers in each cluster.

Plate 3. Desert-Hyacinth

PLATE 4. FAN PALM

PLATE 5. JOSHUA TREE

PLATE 6. SPANISH BAYONET

FAN PALM (*Washingtonia filifera*), plate 4, is California's only native palm. As used in long rows along the sides of a road, it does not seem to me very decorative, but as it grows on the desert, in clumps of various heights and irregularly placed, it is most attractive. Distributed about seeps, springs, and moist places from the southernmost Mojave Desert to Lower California and Arizona, it adds much to the landscape. It attains a height of perhaps eighty feet and bears long open inflorescences with small flowers and, later, hard fruits that are sweet and datelike in flavor.

California has four species of the genus *Yucca*, three of which are shown here and one in the companion volume, *California Spring Wildflowers.* JOSHUA TREE (*Yucca brevifolia*), plate 5, is an amazing plant when thought of as a relative of the Lily, with its tall branched woody habit and overlapping harsh leaves. A close look at the flowers, however, shows a quite lilylike aspect with three outer and three inner fleshy petallike segments about two inches long. Found on much of the Mojave Desert, it attains a height of forty or more feet and blooms from March to May. Underground suckers may be sent out and start new individuals.

Yucca baccata, sometimes called SPANISH BAYONET, plate 6, is a low plant to perhaps four feet high, with long bluish-green leaves having coarse loose fibers along the edge. The flower cluster is very dense and heavy with a red-purple tinge, the flowers being as much as four inches long. They are followed by large fleshy fruits, four inches or more long, that were eaten by the Indians. More common in the states east of California, this Yucca is found between 3,000 and 4,000 feet, in the mountains of the eastern Mojave Desert and flowers from April to June. The fibers of the leaves have been put to use.

Mohave Yucca (*Yucca schidigera*), plate 7, is much more common than *Y. baccata* in California. It resembles that species in its long leaves with marginal curling fibers, but the leaves are mostly a yellow-green and it has trunks up to several feet high. The flowers are cream or with purplish tinge and an inch to two inches long. The fruit is dryer and more of a capsule. This species is abundant on dry rocky slopes and mesas below 7,000 feet, in most of the desert east to Nevada and Arizona and extends into coastal valleys from San Bernardino County south. Flowers appear in April and May.

Plate 7. Mohave Yucca

California has had three species of Century Plant or Maguey or Agave, which belong to the genus *Agave*. One of these is a plant of limestones in the eastern Mojave Desert and has very narrow panicles; it is *A. utahensis* (page 91). A second is coastal with broad deep-green leaves bearing red marginal prickles and growing originally from the San Diego region into Lower California. It is a winter bloomer. The third species is the Desert Agave (*A. deserti*), plates 8 and 9, which has grayish leaves with pale prickles. The trunk is subterranean, but the flower stalk has open branching and becomes fifteen or more feet tall. The yellow flowers are one and one-half to two inches long, followed by dry seed-pods one to two inches long. It can be found in washes and on dry slopes below 5,000 feet, along the western edge of the Colorado Desert and sparingly in the Providence, Old Dad, Granite, and Whipple mountains. It flowers from May to July. Indians here and in Mexico used century plants in many ways: they roasted the white base of the plant for food, fermented the sap from the young flower stalks, manufactured sandals and rope from the fibers, and constructed roofs from the dried flower stalks and overlapping shinglelike leaves.

Plate 8. Century Plant

Plate 9. Century Plant

PLATE 10. DESERT MARIPOSA

PLATE 11. WILD-BUCKWHEAT

PLATE 12. MISTLETOE

Of all the desert wildflowers probably the most colorful is the DESERT MARIPOSA (*Calochortus Kennedyi*), plate 10. The stem may be only a few inches high or up to a foot or more; the lower leaves are slightly grayish, with a length of four to eight inches. The flowers are one to six in a cluster, vermilion to orange, often with brown-purple spots near the base, the petals being one to two inches long. This brilliantly colored flower is found in heavy soil of open or bushy places, between 2,000 and 6,500 feet, from Inyo County to San Bernardino County. It blooms from April to June. In the more eastern parts of the Mojave Desert the vermilion is largely replaced by an orange or even yellow-flowered type.

WILD-BUCKWHEAT (*Eriogonum fasciculatum* var. *polifolium*), plate 11, is a usually many-stemmed shrub one to two feet tall, with numerous grayish leaves about one-half inch long. Common on dry slopes below 7,000 feet, this form of Buckwheat extends over both deserts into the San Joaquin Valley and inner coastal southern California and east to Utah. It blooms in April and May. In the extreme southern Mojave Desert and in the Eagle Mountains it is largely replaced by a low spreading green-leaved form, variety *flavoviride*. See also page 91.

A MISTLETOE (*Phoradendron juniperinum*), plate 12, of the eastern Mojave Desert grows on the Utah Juniper. The leaves are reduced to scales, the plant is smooth and not hairy, with rather stout stems. It bears solitary spikes with a single joint which produces only a few flowers. Much like it in the small scalelike leaves, but with grayish or reddish, more or less pubescent joints is *P. californicum,* a common parasite on Catclaw, Tamarix, Palo Verde, Creosote Bush, and other desert shrubs. It is conspicuous with its reddish berries.

Winter Fat (*Eurotia lanata*), plate 13, is an erect shrub, one to three feet tall, white-woolly throughout and with slender entire leaves. They consist either of a four-lobed calyx with four stamens or of a pair of bracts with a pistil. The flowers, however, are so numerous that they make great tufts with silvery or rusty hairs. Common on flats or rocky mesas above 2,000 feet, Winter Fat is found from the desert to western Kern County, north on the east side of the Sierra Nevada and Cascade Range to eastern Washington and into the Rocky Mountains and Texas. It blooms in the spring. See page 64.

Plate 13. Winter Fat

Hop-Sage (*Grayia spinosa*), plate 14, is a gray-green, often spinose shrub of the same family as Winter Fat. It has fleshier leaves and the two bracts around the female flower are united to form a reticulate sac, which becomes whitish to reddish and almost half an inch long. Rare in the western Colorado Desert, Hop-Sage is common in the Mojave Desert, between 2,500 and 7,500 feet, and ranges north to Lassen and Siskiyou counties, then to eastern Washington and to Wyoming and Arizona. The flowers come from March to June.

Plate 14. Hop-Sage

In the Four-O'Clock Family in addition to Sand-Verbena, page 14, is the Four-O'Clock, page 64, of which our largest native species is *Mirabilis Froebelii*, plate 15. From a thick woody tuberous root it forms sprawling masses that are hairy-viscid throughout, with broad leaves one to three inches long and rose-purple to deep pinkish flowers one to two inches long, of which there are several in a common green involucre. Found in dry stony places below 6,500 feet, this Four-O'Clock ranges from the deserts west to San Luis Obispo and San Diego counties and on the desert north to Mono County. It flowers from April to August. Occasional plants are quite smooth and not sticky.

Plate 15. Four-O'Clock

Plate 16. Barberry

Plate 17. Prickly Poppy

Plate 18. California Poppy

Barberry or Mahonia (*Berberis haematocarpa*), plate 16, is a stiff gray-green shrub with harsh spine-toothed leaves having three to seven leaflets. The flowers are yellow, about one-sixth inch long, see page 92. The juicy purplish-red berries, about one-third of an inch in diameter, yield a dark purple dye and the wood a deep yellow dye, both of which the Indians used. Flowering in May and June, this species is found in dry rocky places between 4,500 and 5,500 feet in mountains of the eastern Mojave Desert. It ranges to Texas and northern Mexico.

Prickly Poppy (*Argemone corymbosa*), plate 17, is a spiny perennial with orange-colored sap and stems two to three feet tall. The leaves are lobed and prickly. The white crinkly flowers are two to three inches in diameter and have numerous yellow stamens. The spiny capsule is to about one inch in length. A beautiful plant, yet one not easily handled because of its armature, it blooms in April and May and is found in dry places between 1,400 and 3,500 feet in the Mojave Desert. As in other poppies (see page 66), it sheds its sepals when the flowers open.

California Poppy (*Eschscholzia Parishii*), plate 18, is an annual. The sepals are grown together into a cap that lifts off when the flower opens. The yellow petals reach the length of about one inch; the capsule two to three inches. This species is found on rocky slopes below 4,000 feet, from the southern Mojave Desert to the Colorado Desert and blooms in March and April. Another desert species, also openly branched, but with much smaller flowers, is *E. minutiflora*. A more distinct one is *E. glyptosperma* which is tufted, the leaves being near the base of the plant and the naked stems ending each in a solitary flower. It is found mostly on the Mojave Desert.

A member of the Mustard Family and with the characteristic four petals (see pages 67, 93), is PRINCE'S PLUME (*Stanleya pinnata*), plate 19. A plant more or less woody at the base, it attains a height of three or four feet and has much divided leaves two to eight inches long and elongate terminal racemes of yellow flowers, each about one-half inch long and with exserted stamens. The seed-pod is linear, one to three inches in length. Apparently largely in selenium-bearing soils of desert washes and slopes at 1,000 to 5,000 feet, it occurs from the north base of the Santa Rosa Mountains north to Inyo County and west to Cuyama Valley toward the coast. Eastward it ranges to North Dakota, Kansas, and Texas. Flowering is from April to September. Another species is *S. elata* of the Death Valley region, wtih entire undivided leaves.

PLATE 19. PRINCE'S PLUME

In the same Mustard Family is PEPPERGRASS (*Lepidium flavum*), plate 20, an almost prostrate annual with yellow flowers and short rounded seed-pods (see also page 67). It is found generally in washes and flats which are somewhat alkaline, below 4,500 feet, from Inyo County to Imperial County and to Nevada and Lower California. It blooms from March to May.

PLATE 20. PEPPERGRASS

Related to the Mustard Family and resembling it in its four petals is the Caper Family (see page 94). One of the most common desert representatives is BLADDERPOD (*Isomeris arborea*), plate 21, an ill-scented shrub two to several feet high, widely branched, and having leaves with three leaflets. The yellow flowers are attractive, about one-half inch long. The pods are inflated, some forms rather narrow and some almost round. The plant is common in somewhat salty places, like bluffs along the seacoast, interior valleys, and deserts below 4,000 feet.

PLATE 21. BLADDERPOD

PLATE 22. LIVE-FOREVER

PLATE 23. APACHE PLUME

PLATE 24. CLIFF-ROSE

LIVE-FOREVER (*Dudleya saxosa*), plate 22, see page 16, varies from pale green to shades of bronze and red, both in its fleshy leaves and stems. The sepals are red, the petals proper yellowish but becoming more or less reddish in age. A form called subspecies *aloides* with practically no red occurs also, especially on desert slopes of the Laguna and San Jacinto mountains and in San Bernardino County mountains. The true *D. saxosa* is in the Panamint Mountains west of Death Valley, at between 3,000 and 7,000 feet.

APACHE PLUME (*Fallugia paradoxa*), plate 23, is a shrub of the Rose Family resembling Quinine Bush (plate 24) and Antelope Brush (page 95). It is distinguished from both in having several pistils in a white flower that is about one-half inch long and borne on the end of a long bare stem. The leaves are dissected into linear divisions with revolute margins and the feathery styles become an inch or longer in fruit. Apache Plume grows on dry rocky slopes between 4,000 and 5,600 feet in the mountains of the eastern Mojave Desert and eastward to Nevada, Texas, and Mexico. The flowers come in May and June.

QUININE BUSH OR CLIFF-ROSE (*Cowania mexicana* var. *Stansburiana*), plate 24, has gland-dotted leaves and solitary cream-colored flowers at the ends of short branches. The petals are about one-third inch long. This shrub grows on dry slopes and in canyons at 4,000 to 8,000 feet, from the White Mountains to the Providence Mountains and thence to Colorado and Mexico. It flowers from April to July. The styles are one to two inches long in fruit. Both it and Apache Plume are frequently browsed on.

DESERT APRICOT (*Prunus Fremontii*), plate 25, is a rigidly branched deciduous shrub (see page 69), which blooms very early in the spring and has sweet plumlike or cherrylike blossoms. The flower is white and about one-half inch across and results in a yellowish fruit about one-half inch long with a large stone. It is found in rocky places along the western edge of the Colorado Desert, as on the Banner Grade, in flower in February and March.

PLATE 25. DESERT APRICOT

FAIRY-DUSTER or MESQUITILLA (*Calliandra eriophylla*), plate 26, is a densely branched, more or less spreading little shrub a foot or so high and with gray pubescent twigs. The leaves are divided into small leaflets. The rose to reddish-purple flowers are produced in dense heads on the ends of short stems and are made most noticeable by the tufts of reddish stamens that may be almost an inch long. The seed pods are typical legumes (pealike pods), two to three inches in length, flat, silvery-pubescent with dark red margins. This handsome little shrub is found in sandy washes and gullies below 1,000 feet, from Imperial and eastern San Diego counties to Texas, Lower California, and central Mexico. The flowers come mostly in February and March.

PLATE 26. FAIRY DUSTER

Resembling Fairy Duster in the flowers not being pealike, but with them yellow, is our common MESQUITE (*Prosopis juliflora* var. *Torreyana*), plate 27. It is a large shrub or low tree with crooked arched branches and forked leaves, each division then pinnate into many small leaflets. The flowers are minute and arranged in spikes two to two and one-half inches long. The pods are two to six inches in length. Common in washes and low places below 5,000 feet, Mesquite is in both deserts and ranges into the upper San Joaquin Valley and Cuyama Valley of Santa Barbara County, as well as other interior valleys of the coastal drainage, also eastward into Mexico. Flowers are from April to June.

PLATE 27. MESQUITE

PLATE 28. DESERT SENNA

PLATE 29. PALO VERDE

PLATE 30. SCARLET LOCOWEED

Another member of the Pea Family, but with more highly developed flowers than in Mesquite is DESERT SENNA (*Cassia armata*), plate 28. A much-branched rounded shrub up to three or four feet high, its numerous yellow-green stems are leafless much of the year. The leaves have one to four pairs of leaflets. Flowers are almost one-half inch long and the yellowish spongy pods an inch to an inch and one-half long. This shrub is common in sandy washes and open places below 3,700 feet, over much of the desert and into Nevada and Arizona. The fragrant flowers come in April and May.

Palo Verde means "green tree" and is an appropriate name for a number of shrubs or small trees related to Senna (page 36). Our most common California species is *Cercidium floridum*, plate 29. The smooth blue-green bark and slender branchlets, mostly spiny and leafless much of the year, are characteristic. The yellow flowers are very numerous and appear from March to May. Palo Verde is often abundant in washes and low sandy places below 1,200 feet, on the Colorado Desert and eastward. Another species of the Whipple Mountains has four to eight pairs of leaflets, while the above has one to three pairs.

SCARLET LOCOWEED (*Astragalus coccineus*), plate 30, is one of the most vividly colored members of the huge genus *Astragalus*. Almost stemless, it makes small clumps or tufts about six to ten inches across, with grayish silvery leaves and many bright red flowers. This gorgeous little desert plant is found in canyons and on gravelly ridges between 2,100 and 7,000 feet. Its distribution is from Owens Valley to Death Valley and south along the edge of the deserts to northern Lower California and in Arizona.

DYEWEED (*Dalea Emoryi*), plate 31, is a densely branched shrub one to three or more feet high with a feltlike wool and is sprinkled with orange glands (see also page 83). The leaves are divided into five to seven leaflets; the calyx is rusty-hairy and the corolla is rose to purplish and about one-sixth inch long. This species grows in dry open places below 1,000 feet, from the Colorado Desert to Lower California and Sonora, blooming from March to May. It gets its common name from the fact that the flower heads can be made to yield a yellowish dye.

LOCOWEED (see page 17) is an immense group and in California, as in most of the western states, is a very complex one. *Astragalus lentiginosus* var. *Fremontii*, plate 32, is one of the common desert locoweeds, perennial, with pinnately compound leaves (having the leaflets arranged like the parts of a feather), and loose racemes of purplish flowers. The seed-pods are not much like the ordinary legume, but very much inflated and papery-membranous. Common on the Mojave Desert below 6,500 feet, it ranges from the east slope of the Sierra Nevada and from the White Mountains to southern Nevada. In the Colorado Desert it is largely replaced by other varieties.

Probably the most dominant and widespread shrub of our southwestern deserts is CREOSOTE BUSH or GREASEWOOD (*Larrea divaricata*), plate 33 (see page 97). Almost always rather scattered, since its roots spread out to some distance not far below the surface of the soil where they can absorb what moisture is available after rains, Creosote Bush has a range from Owens Valley to southwestern Utah, south to San Luis Potosi and east to beyond El Paso. Furthermore, it is found in the deserts of Chile and Argentina. Its yellow petals are twisted a bit and remind one of the parts of a windmill. Because of its strong flavor and resinous sap it is not browsed on as much as are many desert shrubs.

PLATE 31. DYEWEED

PLATE 32. LOCOWEED

PLATE 33. CREOSOTE BUSH

PLATE 34. FREMONTIA

PLATE 35. ROCK-NETTLE

PLATE 36. BLAZING STAR

FLANNEL BUSH OR FREMONTIA (*F. californica*), plate 34, is mostly a plant of the western side or coastal side of the mountains, but it does occur on the desert slopes of the western Colorado Desert and those bordering the western Mojave Desert in sufficient quantity to deserve mention. It is a striking plant, shrubby, low to quite tall, with rounded three-lobed leaves. It bears a profusion of yellow flowers one and one-half to two inches across, which appear largely in May and June, since these desert plants are for the most part from above 3,000 feet where the spring is not early.

ROCK-NETTLE (*Eucnide urens*), plate 35, is a fitting name for a plant not at all related to the true Nettle, but it does have stinging hairs. It is kin to Sandpaper Plant (*Petalonyx*) and Blazing Star (*Mentzelia*), pages 72, 98. It is a rounded bush one to two feet high, very hispid, with coarsely toothed ovate leaves one to two inches long and with cream flowers one to one and one-half inches long. It grows in dry rocky places at 2,000 to 4,500 feet, from the Death Valley region to Utah and Arizona and it flowers from April to June. The stems are straw-colored.

BLAZING STAR is a name given to many different species of *Mentzelia* (see page 98). One of the finest is *M. involucrata* var. *megalantha,* plate 36, an erect bushy annual to about one foot high and with stout pubescent stems and coarsely toothed leaves. The petals are pale cream, satiny, with reddish veins, and in this variety are one and one-half to two and one-half inches long. In *M. involucrata* itself they are not over one inch long. The species and the variety are found in sandy, gravelly, or rocky places below 4,500 feet and bloom from January to May.

Of the cacti one of the common genera in the whole southwestern part of the United States is *Opuntia,* with jointed stems. The joints may be cylindrical as in the Cholla group or flattened in the Prickly-Pear group. In the JUMPING CHOLLA or BALL CHOLLA (*Opuntia Bigelovii*), plate 37, we have an erect plant, usually with a single trunk and a close terminal group above of short lateral branches that are densely set with straw-colored spines, while those on the main trunk become quite black. The flowers are yellow to pale green, over an inch long above the ovary. This species is locally common on fans and lower slopes generally below 3,000 feet, on the Colorado Desert and on the southern Mojave Desert and to Nevada and Arizona. Flowering is mostly in April.

PLATE 37. JUMPING CHOLLA

Another Cholla is SILVER CHOLLA or GOLDEN CHOLLA (*Opuntia echinocarpa*), plate 36, intricately branched, about two to five feet tall, and with rather long joints having tubercles on the surface of the joints less than twice as long as wide. The silvery to golden spines are about one inch long, the pale greenish-yellow petals about an inch long. The species is found in dry well-drained places below 6,000 feet, from Mono County to Lower California, Utah, and Arizona, and blooms from April to May.

PLATE 38. SILVER CHOLLA

The BUCKHORN CHOLLA (*Opuntia acanthocarpa*), plate 39, is three to six feet tall, openly branched, and has surface tubercles an inch or more long (more than twice as long as wide). The straw-colored spines are about an inch long. The petals are red to yellow or greenish-yellow and one inch to an inch and one-half long. Found on dry mesas and slopes below 4,500 feet, it occurs only east of Twentynine Palms and Imperial Valley, ranging to Utah and Sonora. Flowers come in May and June.

PLATE 39. BUCKHORN CHOLLA

PLATE 40. BEAVER TAIL

PLATE 41. OLD MAN CACTUS

PLATE 42. MOJAVE PRICKLY-PEAR

BEAVER TAIL (*Opuntia basilaris*), plate 40, has flat joints and is spineless, but has wicked long sharp hairs or glochids so that it cannot easily be handled. The joints are often beautifully colored grayish to lavender or purplish and the flowers tend to be rose to orchid. They are clustered at the upper ends of the joints and have petals to an inch and one-half long. Hence they are very showy, appearing from March to June. Beaver Tail has several forms and habits of growth. It frequents dry benches and fans below 6,000 feet and reaches Arizona and Utah on the east and interior valleys of the coastal drainage on the west.

OLD MAN and PRICKLY-PEAR are common names for *Opuntia erinacea,* plate 41, while Grizzly Bear Cactus is used for a form with flexible spines three to eight inches long (var. *ursina*). The plants are low and form creeping masses. The petals are yellow, sometimes red in age, and about one inch long. Found in various forms, the species occurs on dry gravelly and rocky slopes below 4,500 feet, sometimes higher, and ranges from the Santa Rosa Mountains to the east slope of the Sierra Nevada and the White Mountains and eastward. It blooms mostly in May and June.

MOJAVE PRICKLY-PEAR (*Opuntia mojavensis*), plate 42, is a prostrate plant with stems to four or five feet long and large flat often erect joints to about one foot long. The spine-clusters are remote, one to two inches in length, the spines being red-brown at base, white to yellow or yellow-brown toward the tip. The petals are pale yellow to orange, from about one to two inches long. The red-purple fruits are conspicuous. Found on dry slopes and in washes between 4,000 and 5,000 feet, it occurs in the eastern Mojave Desert and flowers in May and June.

Hedgehog Cactus, "Torch Cactus" (*Echinocereus Engelmannii*), plate 43, is named for the first important cactus student of this country, Dr. George Engelmann, 1809-1884, of St. Louis. This plant has one to few cylindrical stems, to about one foot high, each with ten to thirteen ribs. The flowers are crimson-magenta to paler, one and one-half to three inches long. It is common on gravelly slopes and benches below 7,200 feet, in both our deserts, from the White Mountains to Lower California and eastward to Utah and Sonora. It flowers in April and May.

Plate 43. Hedgehog Cactus

Mound Cactus is another *Echinocereus*, namely *E. mojavensis*, plate 44. It has many stems, up to sixty or more in fact, and commonly forms veritable mounds of pale green stems. The flowers are dull scarlet, two to almost three inches long and often do not open out completely when in bloom. It can be found on rocky slopes between 3,000 and 7,000 feet, from the San Bernardino Mountains to the White and Clark mountains and eastward, flowering from April to June. Because of its color and habit of growth it is one of our most noticeable cacti.

Plate 44. Mound Cactus

Foxtail Cactus (*Mammillaria Alversonii*), plate 45, has one to few short-cylindrical stems, four to eight inches high, with pinwheellike spine clusters that are white or ashy. The flowers are about an inch long, magenta with deeper red midveins and white stigmas. The word Mammillaria comes from the teatlike tubercles and members of the genus are sometimes called Nipple Cactus. This species is found on stony slopes at between 2,000 and 5,000 feet, in the Little San Bernardino Mountains, Eagle Mountains, and Chuckawalla Mountains. The flowers appear in May and June.

Plate 45. Foxtail Cactus

PLATE 46. DESERT-PRIMROSE

PLATE 47. BOTTLE CLEANER

PLATE 48. EVENING-PRIMROSE

The Evening-Primrose Family (see page 99), has the flower built on the plan of four and the ovary below the flower. One of the best known species is DESERT-PRIMROSE, DEVIL'S LANTERN, or LION-IN-A-CAGE (*Oenothera deltoides*), plate 46, a coarse spring or winter annual with several named variants. The flowers open in the early evening and the white petals are one to one and one-half inches long, aging pink. The capsule is a narrow structure one to two inches in length. Common in sandy places below 3,500 feet, this plant, in one form or another, extends over the deserts, into the San Joaquin Valley to the west, to Modoc County to the north, and to Utah, Arizona, and Lower California. Flowers come mostly from March to May. It has four linear lobes to the stigma.

An Evening-Primrose with white flowers and a round stigma is *Oenothera decorticans* var. *condensata,* plate 47. The species is an annual and has several forms varying in shape of petals and thickness of stem and capsule. The dried persistent seed-pods make a cluster resembling a test-tube cleaner, and it is sometimes called BOTTLE CLEANER. On open slopes and plains below 6,000 feet, the species ranges from the deserts to Monterey and San Benito counties on the west and to Utah and Nevada on the east.

EVENING-PRIMROSE (*Oenothera caespitosa* var. *marginata*), plate 48, is a cespitose perennial, stemless or short stemmed, more or less hairy, with narrow lobed leaves and fragrant large flowers that open in the evening. It differs from the Desert-Primrose (plate 46) in its shorter thicker tapering fruit with little bumps or tubercles along it. This plant is occasional on dry mostly stony slopes between 3,000 and 10,000 feet, from the Santa Rosa Mountains to the White Mountains and to Utah and Arizona.

An Evening-Primrose with yellow flowers and a spherical stigma is *Oenothera cardiophylla,* plate 49, sometimes called Heart-leaved-Primrose. It is annual to perennial, usually branched and soft-hairy, one to one and one-half feet high, with somewhat rounded leaves. The petals are pale yellow, aging red, one-third to almost one inch long. The seed-pod is on a short stalk. It is found in desert canyons and on mesas below 5,000 feet and occurs in the Argus and Panamint mountains and in the Colorado Desert, thence to Arizona and Lower California. It is a spring bloomer.

Plate 49. Heart-leaved-Primrose

In the same group with smallish yellow flowers and a club-shaped capsule on a short stalk and with spherical stigma is *Oenothera clavaeformis* var. *Peirsoni,* plate 50, of sandy flats in the Colorado Desert. The leaves are pinnately divided into many segments and the stems are hairy. It blooms in early spring. I know no good common name. White-flowered varieties of *O. clavaeformis* are common over most of the desert and can be distinguished from *O. decorticans* (plate 47) since the latter has the capsule woodier and not stalked.

Plate 50. Oenothera

Catchfly-Gentian (*Eustoma exaltatum*), plate 51, is not a typical desert species, but since it is found in moist places such as Thousand Palms, Palm Springs region, and San Felipe, and since it is a noticeable plant about which the finder might well be curious, it is here included. It is herbaceous, one to two feet tall, smooth, with clasping upper leaves two to three inches long. The deep blue or blue-lavender corolla is an inch or more long. It is in the Gentian Family and ranges from coastal California to Florida and Mexico. The flowers may be found almost throughout the year.

Plate 51. Catchfly-Gentian

Plate 52. Skyrocket

Plate 53. Ipomopsis

Plate 54. Phlox

In the Phlox Family is Scarlet-Gilia or Skyrocket (*Ipomopsis aggregata* ssp. *arizonica*), plate 52, a biennial or short-lived perennial up to a foot high with short dissected leaves. The flowers are bright red, about one inch long, tubular, and arranged in a narrow crowded panicle. This handsome plant is found in dry washes and rocky places mostly between 4,500 and 10,500 feet, in the Providence, New York, Clark, Panamint, and Inyo mountains, and thence ranging to Utah and Arizona. Flowers may be found from May to October. Other forms of this species are much taller plants of different distribution.

Quite different from the species shown in plate 52 is *Ipomopsis tenuifolia*, plate 53, a perennial from a woody base, many-stemmed, to about one foot high, with linear entire or dissected leaves. The flowers are solitary in the upper leaf-axils, tubular-funnelform, to about one inch long. It is found among rocks at 1,500 to 3,500 feet, in southeastern San Diego County as about Campo and Jacumba, and into northern Lower California. Its flowers appear from March to May.

True Phlox can be told from most other members of its family, which it may resemble, by the stamens being unequally placed in the corolla-tube. A desert species of Phlox (*P. Stansburyi*) is shown in plate 54. It is less than one foot high, from a branched root-crown, and has narrow leaves about one inch long. The flowers are rather few, in open terminal clusters, about an inch long and rose to whitish. It grows on dry gravelly slopes and washes, at 5,000 to 9,000 feet, from the east slope of the Sierra Nevada of Mono and Inyo counties, through the White and Inyo mountains, to Nevada, Arizona, and New Mexico. Flowering is from April to June.

DESERT CALICO, see page 21, is *Langloisia Matthewsii,* shown in color in plate 55. The whitish or pinkish flower is characterized by being two-lipped and by its reddish markings. Only a few inches high, this spring annual often occurs in such masses and persists for so long a time that it colors the sandy floor of the desert with pinkish patches after most spring annuals are out of bloom. Originally placed in the genus *Gilia,* it and three other species with spine-tipped teeth form a natural little group now called *Langloisia* by most botanists in honor of a Catholic priest, Father Langlois.

PLATE 55. DESERT CALICO

Two species of *Langloisia* have the corolla lobes alike, thus not forming two lips. One of these is SPOTTED LANGLOISIA (*L. punctata*), plate 56; see also page 85. Here the corolla-lobes are purple-dotted, the whole corolla being slightly less than one inch long. The narrow leaves are finely bristle-toothed. It is found in dry gravelly places mostly below 5,000 feet, from the White Mountains of Inyo County south to the San Bernardino Mountains and to Nevada and Arizona. It flowers in May and June. In my experience it does not form the masses that Desert Calico does.

PLATE 56. SPOTTED LANGLOISIA

In the Phlox Family the genus *Gilia* has the stamens equally inserted in the corolla-tube, see page 84. In plate 57 is shown *Gilia latiflora,* an erect plant with basal leaves much larger than those on the stems. It is an annual that may or may not have cobwebby hairs at the base. The flowers are somewhat less than an inch long, with a slender purple tube, a full throat that may be yellow or white to violet, and white to violet corolla-lobes. GILIA occurs in a number of forms, the typical one being found on sandy flats and washes at 2,500 to 3,600 feet, in the southwestern Mojave Desert, flowering in April and May.

PLATE 57. GILIA

PLATE 58. GILIA

Another GILIA is *G. cana* ssp. *triceps,* plate 58, with the basal leaves much lobed or toothed and cobwebby, the stems to about one foot high, and the inflorescence open and loose. The corolla is from one-third to almost one inch long, with a slender tube expanding abruptly into the throat. The tube is more or less purplish, the lobes pinkish-violet. This form of the species, which is quite polymorphous, occurs on dry slopes and washes at 2,800 to 5,200 feet, from Barstow and Kelso to the Panamint and White mountains. It flowers in April and May.

PLATE 59. LINANTHUS

LINANTHUS (see page 100), is a group in the Phlox Family with leaves opposite (in pairs) and divided palmately into linear segments. A bushy perennial species is *L. Nuttallii,* plate 59. It grows from a woody base, may have many erect stems four to eight inches or more tall, and its leaves are one-half inch or so long. Found in dry rocky or brushy places at 4,000 to 12,000 feet, it is distributed from Humboldt and Trinity counties to Modoc County, then north to Washington and south along the east slope of the Sierra Nevada and other ranges to Lower California and to the Rocky Mountains and Mexico.

PLATE 60. PHACELIA

PHACELIA OR WILD-HELIOTROPE (see page 85), has a very characteristic desert species in *Phacelia crenulata,* plate 60, occurring in two or three forms over most of our desert. Annual, one to two feet high in good seasons, it is very glandular and strongly scented. It is a handsome plant and grows in gravelly and open places, mostly below 5,000 feet. Its sticky glandular material is poisonous to many persons, including your author, and raises a dermatitis like that produced by Poison-Oak.

The WILD-CANTERBURY-BELL (*Phacelia campanularia*), plate 61, likewise produces dermatitis on many people. It is a striking desert annual, up to two or more feet tall, with broad leaves and deep blue flowers over one inch long. It too is not uniform, but has somewhat different forms in the southern and eastern Mojave Desert from the plants in the western Colorado Desert. It is found in dry sandy and gravelly places, flowering from February to May.

PLATE 61. WILD-CANTERBURY-BELL

The VERBENA is represented from the eastern Mojave Desert to Utah and Arizona by a native perennial, *V. Gooddingii*, plate 62. The stems are several, to a foot or so high, hairy, with somewhat divided and lobed leaves and headlike spikes of purplish flowers almost half an inch in diameter and with a tube almost that long. Found in dry canyons and on slopes at 4,000 to 6,500 feet, this species blooms from April to June. Most of our California verbenas are introduced weeds and a so-handsome native is a pleasing contrast.

PLATE 62. VERBENA

THISTLE SAGE (*Salvia carduacea*), plate 63, is an annual, simple or few-stemmed, one to two feet high, with a rosette of basal prickly leaves and the flowers in small heads in an open inflorescence. These have a woolly calyx and a remarkable lavender corolla about one inch long, the lower middle lobe of which is large, fan-shaped, and fringed. The vermilion or brick-red anthers add to the interesting coloration. Thistle Sage is frequent in sandy and gravelly flats below 4,500 feet, across the western Mojave Desert, then to Contra Costa and Stanislaus counties of coastal drainage and through coastal southern California to lower California. It flowers from March to June.

PLATE 63. THISTLE SAGE

PLATE 64. PURPLE SAGE

PLATE 65. BLADDER-SAGE

PLATE 66. PENNYROYAL

Another conspicuous Sage is BLUE SAGE or PURPLE SAGE (*Salvia Dorrii*), plate 64, a low much-branched shrub with scurfy-hoary leaves less than an inch long, purplish or greenish broad bracts around the flower heads, and bright blue flowers about one-half inch long. It occurs in a number of forms varying in leaf shape and size, growing in dry places between 2,500 and 8,000 feet high, from Los Angeles and San Bernardino counties on the western Mojave Desert to Utah and Arizona, Idaho, and northern California (Lassen County). It blooms from May to July.

BLADDER-SAGE (*Salazaria mexicana*), plate 65, like the true sages is a member of the Mint Family, which is characterized by its strongly aromatic qualities (Horehound, Mint, Oregano, Thyme, for example), often square stems, opposite leaves, and strongly two-lipped flowers. For this species see page 87; it is a low shrub with grayish twigs, smallish leaves, and a calyx that becomes inflated, colored, papery, and bladderlike in fruit. The species is found from Inyo County to northern Mexico.

Quite a large California group in the Mint Family is PENNYROYAL or MOUNTAIN-PENNYROYAL, which is of the genus *Monardella* and not the true Pennyroyal of Europe. We have both annual and perennial species; one of the former is *M. exilis,* plate 66, of open flats between 2,000 and 3,500 feet in the western Mojave Desert. It is an erect annual, branched above, bearing lance-shaped leaves an inch or more long, the stems ending in heads subtended by abruptly pointed bracts ending in white tips, although the bracts themselves may be purplish. The white flowers are about one-half inch long and appear in May and June. Most species of *Monardella* have a pleasant odor when crushed.

Ghost Flower (*Mohavea confertiflora*), plate 67, belongs to the same family as Snapdragon and Penstemon (see pages 50, 101). It is a very viscid annual, simple or branched, to about one foot tall. The corolla is closed at the throat and is an inch or more long, with a purple-dotted palate. Found in sandy washes and on gravelly slopes below 3,000 feet, Ghost Flower ranges on the Mojave Desert from the Ord Mountains and Daggett east and south to Nevada, Arizona, and Lower California, blooming from March to April. The fan-shaped corolla-lips and hairy palate are characteristic.

Plate 67. Ghost Flower

The Lesser Mohavea (*M. breviflora*), plate 68, has much the same habit of growth as Ghost Flower and, like it, is densly glandular. The corolla is lemon yellow and somewhat smaller, its palate not being conspicuously dotted. It has a more limited range, from the Death Valley region to western Nevada and northwestern Arizona, blooming also in March and April and growing in much the same kind of places. On the Mojave Desert it ranges as far south as Kelso and Bagdad. Both species are interesting in having only two functional stamens, the others having become abortive.

Plate 68. Lesser Mohavea

In the same family with Mohavea is Penstemon or Beard Tongue; this large genus has five stamens, but only four are functional. Since there are several desert species and they are quite different from each other, it is interesting to see a number of them shown in color. Among the red-flowered ones, fairly common and widespread at elevations up to 8,000 feet, is *Penstemon Eatonii*, plate 69. It is a perennial with few to several stems one to three feet tall and with thickish glabrous leaves one to four inches long. This species is found on desert slopes from the San Bernardino Mountains to Utah and Nevada and blooms from March to July.

Plate 69. Penstemon

PLATE 70. PENSTEMON

PLATE 71. PENSTEMON OR BEARD TONGUE

PLATE 72. PENSTEMON

Another red PENSTEMON is *P. utahensis,* plate 70, of a carmine color and with flowers less than one inch long. It is generally of lower growth than is *P. Eatonii* and the flowers are narrower and glandular, not smooth as in the former. It is of more limited distribution, being occasional between 4,000 and 5,500 feet in rocky places in the New York and Kingston mountains and thence to Utah and Arizona. It flowers in April and May. A rose-lavender or purplish species of hills surrounding Owens Valley is *Penstemon confusus* var. *patens* with a more open inflorescence.

Penstemon Palmeri, plate 71, is a tall species with leaves often sharply toothed and with an inflorescence eight to twenty-four inches long. The abruptly inflated corolla is strongly two-lipped, white or tinged with pink, lavender, or purple, and has prominent colored lines extending into the throat from the lower lip. It is quite fragrant. The sterile stamen is shaggy-bearded and makes it truly a BEARD TONGUE. It is a species of dry rocky gullies, from 4,000 to 6,000 feet, in the mountains west of Death Valley and in the Kingston, Clark, Providence, and New York mountains, ranging also to Utah and Arizona. It flowers in May and June.

Quite a different PENSTEMON is *P. monoensis,* plate 72, with stems to about one foot high, thickly and closely pubescent. The dense flower cluster is glandular-pubescent, the rose-purple or wine-red corolla being about two-thirds of an inch long, tubular-funnelform in shape. It is found in dry stony places at 3,800 to 6,000 feet along the base of the White and Inyo mountains. A somewhat similar species is *P. calcareus* of the Grapevine and Providence mountains, with a light rose to rose-purple flower.

In the same family with Penstemon is the MONKEY FLOWER, a group characterized by a strongly angled calyx. A desert species is *Mimulus Bigelovii,* plate 73. It is a low densely glandular-pubescent annual, with leaves to about one inch long. The flowers are mostly clustered near the tips of the stem-branches. It is common in dry sandy or gravelly washes and canyons below 9,000 feet, on the Mojave and Colorado deserts from Mono County south. Another small desert glandular annual is *M. mohavensis* of the region about Barstow and Victorville. Each corolla is red-purple with a pale margin.

PLATE 73. MONKEY FLOWER

Still in the family with Penstemon is PAINT-BRUSH (*Castilleja chromosa*), plate 74; see also page 22. It is an herbaceous perennial from a woody root-crown, with stems to a foot or more high and having stiff hairs. The narrow leaves are to about one inch long; the upper bracts and calyces have scarlet tips and give most of the color to the flower spike. This species is frequent on dry brushy slopes between 2,000 and 7,000 feet, from the Pinto Mountains in Riverside County northward to eastern Oregon and eastward to Wyoming, Colorado, and New Mexico. Flowers may be found from April to August, but on our desert, only in the early season.

PLATE 74. PAINT-BRUSH

For a discussion of CHUPAROSA see page 23. It is shown here in color, since it is a striking plant of sandy washes and watercourses of the Colorado Desert. It is *Beloperone californica,* plate 75. In the tropics the Acanthus Family to which Chuparosa belongs is very large. The Greeks used the leaves of a Mediterranean species of Acanthus as the model for the decoration of the capital of the Corinthian column.

PLATE 75. CHUPAROSA

Plate 76. Goldenhead

Plate 77. Goldenbush

Plate 78. Desert-Aster

For one species of Goldenhead see page 102. Another, *Acamptopappus Shockleyi,* plate 76, is a rounded shrub, six to eighteen inches high, spinescently branched, minutely rough-hairy; leaves are to about one-half inch long. It differs from *A. sphaerocephalus* in possessing ray-flowers at the edge of the head and in having the heads solitary at the ends of the branches. It is found on plains and washes, between 3,400 and 6,200 feet, from the White Mountains of northern Inyo County to the Clark Mountains of eastern San Bernardino County and into southern Nevada. Flowers appear from April to July.

Goldenbush is discussed on page 103. It (*Haplopappus linearifolius*), plate 77, is a much-branched shrub with crowded linear leaves that are covered with gland-dots and get to be about one and one-half inches long. There are other desert species of *Haplopappus,* a common one being *H. Cooperi,* which is a low flat-topped shrub with almost linear leaves up to one-half inch long. These primary leaves are gland-dotted and have fascicles of smaller leaves in their axils. The florets are mostly tubular and ray-flowers are but little developed. It is one of the common low shrubs at above 2,000 feet, from Lancaster Valley and the Little San Bernardino Mountains northward.

Desert-Aster or Mojave-Aster is figured on page 87, but I cannot refrain from including also a color shot of this lovely plant, plate 78. It has from forty to sixty ray-flowers (the petallike outer ones) in each head which vary from blue-violet to lavender to nearly white and are up to one inch long. Common from Riverside County to Inyo County, it adds much to the color of rocky places. Along the northern and western edges of the Colorado Desert it is replaced by two other closely related species, all three agreeing in having large showy heads.

A Desert Daisy is *Erigeron pumilus* ssp. *concinnoides,* plate 79, a perennial with a taproot and a many-headed caudex at its summit. The leaves are rather harsh, up to three or four inches long. The terminal flower-heads are hemispheric, with fifty to one hundred rays or petallike florets which can be blue, pink, or white. Found on dry gravelly slopes at 4,000 to 6,000 feet, this plant occurs in the mountains of the eastern Mojave Desert of Inyo and San Bernardino counties, blooming in May and June.

Plate 79. Desert Daisy

Viguiera (see page 105), is represented also by *V. multiflora* var. *nevadensis,* plate 80, a perennial herb with several slender erect stems with gland-dotted leaves one to two inches long. The heads are few, in loose clusters, and about an inch or more across. It is found in canyons at 4,000 to 7,500 feet, in mountain ranges of Inyo and northeastern San Bernardino counties and reaches also Utah and Arizona. It blooms from May to September, being able to keep going until summer rains renew its period of activity.

Plate 80. Viguiera

Panamint-Daisy (*Enceliopsis argophylla* var. *grandiflora*), plate 81, is another Sunflower relative and a stunning one. With a stout woody taproot, it is perennial and has large silvery leaves on long petioles. The flower-heads are borne on long naked stems and have a central disk up to two inches broad and around this the ray-flowers, which are about an inch and one-half long. It is a rare plant, being found in rocky or clayey places that are more or less alkaline and occur between 1,200 and 4,000 feet on the west side of the Panamint Mountains. It blooms from April to June. A related species, *E. nudicaulis,* has the leaves less sharply pointed and the heads of flowers smaller.

Plate 81. Panamint-Daisy

Plate 82. Coreopsis

Plate 83. Cheese Bush

Plate 84. Layia or White Tidy-Tip

Coreopsis (see page 105), is shown here as *C. calliopsidea,* plate 82, a stout almost fleshy annual, each stem ending in a single showy head. The leaves are two to three inches long and with narrow linear lobes. The heads are one to three inches across, golden. It is found in the western Mojave Desert of Kern, Los Angeles, and San Bernardino counties and into the coastal drainage from Los Angeles County to Alameda County. It has flowers from March to May.

Cheese Bush or Bush-Hops (*Hymenoclea Salsola*), plate 83, belongs to the Sunflower Family, but has its heads so reduced as not to resemble most members of that family. It is a shrub with narrow resinous leaves one to two inches long and with male and female flowers in separate heads but on the same plant. Each female or pistillate head has one floret surrounded by an involucre and resembling in some ways Cheeseweed of the Mallow Family. The staminate or male occur above the pistillate. The illustrations show several of the latter. Cheese Bush is common in sandy washes and rocky places below 6,000 feet and is widespread over the desert and into dryer parts to the west, as in the upper part of the San Joaquin Valley.

Tidy-Tip is the name given to a member of the Sunflower Family, *Layia,* in which the yellow ray-flowers have white tips. Another Layia (*L. glandulosa*), plate 84, is sometimes called White Tidy-Tip. It is a glandular annual, one to two feet high. The heads are an inch or more across. This Layia is common in sandy soil at below 7,000 feet, over much of our desert and in areas of the coastal drainage as far north as Contra Costa County and south to Lower California. It also ranges to eastern Washington, Idaho, and Utah, blooming from March to June.

PAPERFLOWER is figured on page 106 and here is offered in color, plate 85, (*Psilostrophe Cooperi*). Dr. J. G. Cooper, for whom it was named, lived from 1831 to 1902 and was a physiologist and ornithologist who discovered many of our desert plants, having botanized in the Mojave Desert in the 1860's.

PLATE 85. PAPERFLOWER

A tufted perennial from a branched woody base is *Hymenoxys acaulis* var. *arizonica*, plate 86. The leaves are basal, gray-silky, one to two inches long. The heads are solitary and made up of bright yellow florets, both the tubular disk-florets and the marginal rays. These heads are perhaps an inch across. The plant is found apparently largely on limestone at between 4,000 and 8,000 feet, in the Chuckawalla, Providence, New York, and Clark mountains of our eastern desert region and then ranges to Colorado. It flowers from April to June. Its relationship has apparently not been well understood, since it has been referred to eight different genera.

PLATE 86. HYMENOXYS

Some species of ERIOPHYLLUM are called Golden-Yarrow, but they seem largely to be perennial or subshrubby. I am put to it to find a common name for *Eriophyllum Wallacei*, plate 87, a very common little desert annual. Persistently woolly, usually branched, it varies from little tufts to mats. The leaves are three-lobed at the tip, the heads scattered, each with five to ten golden or yellow rays, although in one form on the western edge of the Colorado Desert these may be reddish or purplish. The species is common on sandy flats and fans below 5,000 feet, from Mono County to Lower California, ranging eastward to Utah and Arizona. It is a spring bloomer.

PLATE 87. ERIOPHYLLUM

Plate 88. Desert-Marigold

Plate 89. Groundsel

Plate 90. Desert Velvet

Desert-Marigold, see also page 107, has three species in California. The one shown here, *Baileya multiradiata,* plate 88, is a biennial or perennial, leafy only below the middle, and with large heads on naked stems. The ray-flowers are numerous and about one-half inch long, so that the heads have a diameter of more than an inch. It is an attractive plant and will continue to bloom for many months if it has a little moisture, making a good garden plant. It is found wild on sandy plains and rocky slopes, at from 2,000 to 5,200 feet, from the eastern Mojave Desert to Utah and Texas. There it flowers in the spring and more sparingly in the fall.

Groundsel, see page 108, is shown in plate 89, as *Senecio stygius.* It is an unusually attractive species of a large genus with some species arborescent, others shrubby, some found above timberline. This one from the eastern Mojave Desert is quite without hair except for little tufts of wool in the axils of the leaves. It is found with Joshua Tree and among Pinyon and Juniper in the mountains, blooming from April to May.

Desert Velvet is a name given to a compact round and rather flat plant which is mostly an annual and covered with white wool, *Psathyrotes ramosissima,* plate 90. It has a strong turpentinelike odor and numerous thick, coarsely and irregularly toothed leaves. The heads are about one-fourth inch high with tubular yellow to purplish corollas. The plant is common on dry hard soil of flats and ledges, largely at below 3,000 feet, on both deserts, especially in the eastern part. It ranges to Utah, Arizona, and northwestern Mexico, flowering mostly from March to June but sometimes in the winter.

For one species of TETRADYMIA or COTTON THORN see page 109. Here is presented *T. axillaris,* plate 91, a rigidly branched shrub, densely white-woolly, to about three or four feet high and with the primary leaves modified into inch-long slender spines. The heads are a clear yellow, each with a half-dozen florets. It is common on dry slopes and flats, at 2,000 to 6,400 feet, from the Mojave Desert north to Mono County and east to Utah and Arizona. Flowers come in the spring.

There are not many species of THISTLE on the desert, but one, *Cirsium neomexicanum,* plate 92, occurs rather widely if soemwhat sparingly. It is a tall white-woolly biennial with strongly spinose leaves to one and one-half feet long. The heads are about two inches broad, with a sparsely woolly subglobose involucre surrounding the very slender numerous whitish flowers. This species is found in dry rocky places between 3,500 and 6,000 feet, from the mountains of the eastern Mojave Desert to Colorado and New Mexico, and it blooms in April and May. Another species is *Cirsium mohavense* of moist alkaline places, such as those near Rosamond in Antelope Valley. In it the middle bracts of the involucre have a conspicuous glutinous ridge.

DESERT-DANDELION (*Malacothrix glabrata*), plate 93, like the true Dandelion, has all the florets of the head strap-shaped or petallike. It is an annual, usually many-stemmed, each branch bearing a few heads of fragrant pale yellow flowers. The plants are quite smooth and hairless and are abundant in dry sandy plains and washes below 6,000 feet, on both deserts and sometimes in the hot interior valleys of the coastal drainage from San Diego County to Santa Barbara County. Eastward it ranges to Idaho and Arizona.

PLATE 91. COTTON THORN

PLATE 92. THISTLE

PLATE 93. DESERT-DANDELION

PLATE 94. YELLOW SAUCERS

PLATE 95. SCALE BUD

PLATE 96. TACKSTEM

Another species of *Malacothrix* is YELLOW SAUCERS (*M. sonchoides*), plate 94. It too is an annual; however, its leaves are not dissected into filiform divisions but have callus-tipped lobes. The bright yellow heads are about one inch across. It is occasional in open somewhat sandy places, at between 2,000 and 5,000 feet, in the western and northern parts of the Mojave Desert and ranges north to Modoc County and east to Nebraska. It blooms from April to June.

The only common name that I have found for *Anisocoma acaulis,* plate 95, is SCALE BUD. It is a low annual with a rosette of pinnately parted or toothed leaves and several ascending one-headed stems. The involucre of these heads is handsome, each bract usually being edged with red and having reddish dots. The whole involucre is so neat and well made that it attracts attention. The florets are all strap-shaped, pale yellow, and the head opens only in sunshine. The species is common in washes and sandy places above 2,000 feet and is found on both deserts, blooming from April to June.

Another member of the Dandelionlike group is TACKSTEM (*Calycoseris Wrightii*), plate 96. It is annual and can be distinguished from most other similar plants on the desert by the tack-shaped glands in the upper parts of the plant. The leaves are pinnately parted into short linear lobes or the uppermost are entire. The rays are white with rose or purplish dots or streaks on the back and may be almost an inch long. It is occasional from Death Valley to the eastern Mojave Desert and the western edge of the Colorado Desert. A similar species in having tack-shaped glands, but with yellow flowers, is *Calycoseris Parryi.* Both bloom in the spring.

FLOWERS WHITE TO PALE CREAM OR PALE PINK OR GREENISH

Section Three

Perhaps only a botanist would think of grasses as having flowers, since he knows that a flower does not necessarily have showy sepals and petals, but does have the essential stamens and pistils (see p. 3). At any rate, let us put in our desert flowering plants two grasses so conspicuous and widespread that surely we want to know their names. The first of these, GALLETA GRASS (*Hilaria rigida*), figure 42, is a rather heavy-stemmed perennial with woody spreading rhizomes forming large open erect grayish-hairy clumps two or more feet tall. The terminal spikes are two to three inches long. This grass occurs mostly in sandy places below 4,000 feet, throughout our deserts and to the immediate east. It is said to be the most valuable forage grass of the desert.

Another common grass is INDIAN MOUNTAIN-RICE or SAND BUNCHGRASS or just RICEGRASS (*Oryzopsis hymenoides*), figure 43, also at dry sandy places, but ascending on the east slopes of the Sierra Nevada and in the Inyo-White Mountains to 10,000 feet. Its range extends from our deserts to British Columbia, Manitoba, Texas, and northern Mexico. Its florets are borne in open panicles on filiform wiry branches and are formed from April to July.

The desert does not seem a likely place for lilies, but quite a few plants of the lily and related families are found in extremely dry parts of the earth, where they carry on by means of their deeply buried bulbs that can store up water and food and remain dormant during dry times. Among such is the MARIPOSA-LILY or SEGO-LILY (*Calochortus Nuttallii*), figure 44, having white petals with a lilac tinge and often a colored spot above the basal gland. It is found in desert mountains between 5,000 and 9,000 feet from Inyo County north,

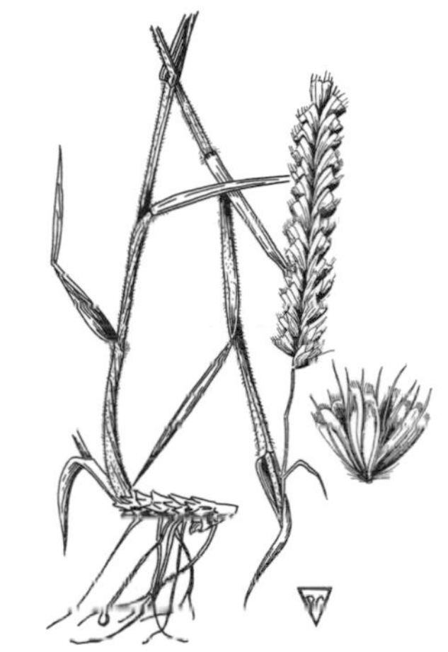

FIGURE 42. GALLETA GRASS

FIGURE 43. INDIAN MOUNTAIN-RICE

FIGURE 44. MARIPOSA-LILY

FIGURE 45. DESERT ZYGADENE

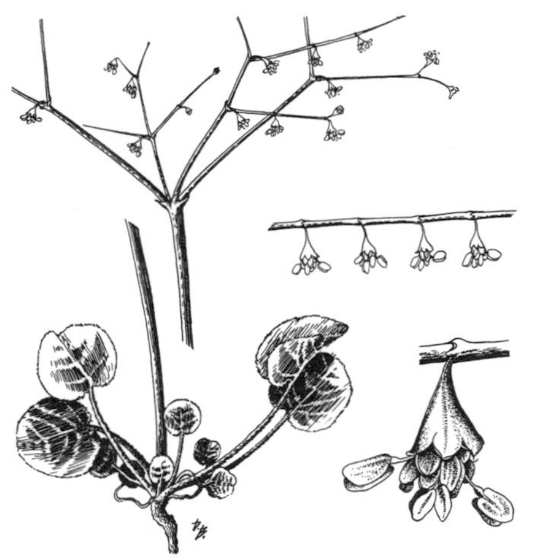
FIGURE 46. SKELETON WEED

FIGURE 47. SAUCER PLANT

and blooms from May to July. For other species, see page 30.

Another member of the Lily Family is the DESERT ZYGADENE (*Zigadenus brevibracteatus*), figure 45, a bulbous plant with a few basal leaves that can attain a height of almost a foot. From between them grows a branching flower stalk to a height of about a foot to a foot and one-half. The creamy-white flowers are six-parted and approximately an inch across. It is found in open flats from the extreme northwestern Colorado Desert through the southern and western Mojave Desert into eastern San Luis Obispo County. It flowers in April and May.

Perhaps no group of plants has more species in our California deserts than the Wild-Buckwheat, an example of which is SKELETON WEED (*Eriogonum deflexum*), figure 46. Many buckwheats are annuals, some perennials, some even shrubs. Belonging to the same family as cultivated Rhubarb, they are characterized by flowers built on the plan of three, with three outer sepallike parts, and three inner petallike ones. Flower color varies greatly (see pages 30 and 91), but our present example is whitish, sometimes pink. It is a widely spreading annual with pendent flowers and is common in washes and on adjacent slopes below 7,000 feet through much of our California desert. Flowers usually bloom from May to October.

In the same family with Wild-Buckwheat, and with its flower built on much the same plan of three, is SAUCER PLANT (*Oxytheca perfoliata*), figure 47. It is a spreading annual with more or less horizontal branches up to a foot long, but the conspicuous feature is the presence of fused bracts that constitute a series of cup-shaped discs about one-half inch across. This species occurs in sandy or gravelly places, at 2,400 to 6,000 feet, from the Mojave Desert to Lassen County and to Arizona.

The TUFTED MISTLETOE (*Phoradendron Bolleanum* var. *densum*), figure 48, is a parasite on California Juniper, hence is found in the western parts of the Colorado and Mojave deserts, mostly below 5,000 feet. The leaves and stems are yellowish-green, the berries straw-colored. It is much-branched, dense, in tufts four to eight inches long. On the more eastern Utah Juniper is a different Mistletoe (*Phoradendron juniperinum*), with scale-like leaves, the plants being hairless when young and with white berries (see page 30). A third Mistletoe, also with scalelike leaves, but with grayish stems finely hairy when young and mostly reddish berries is *P. californicum*. It grows on Catclaw, Mesquite, Creosote Bush, and Palo Verde.

FIGURE 48. TUFTED MISTLETOE

RUSSIAN-THISTLE OR TUMBLEWEED (*Salsola Kali*), figure 49, is a weed that has become tremendously common in the West. Native of Eurasia, it was introduced many years ago. In the spring the young growth is fleshy and readily eaten by cattle, but as it matures, its spine-tipped scalelike leaves become very sharp and its seeds develop without interference. When ripe, it rounds up, breaks loose at the root, and is whirled across the desert plains by every strong wind, scattering its seeds as it travels.

FIGURE 49. RUSSIAN-THISTLE

In the same family (Pigweed or Goosefoot Family) is the large genus *Atriplex* characterized by the pistillate flower being borne between a pair of more or less united bracts, as seen in the upper drawing of figure 50, DESERT-HOLLY (*Atriplex hymenelytra*). The leaves of these atriplices or shadscales are covered with a whitish or grayish indument of dry air-sacs that grow out from the epidermal

FIGURE 50. DESERT-HOLLY

Figure 51. Winter Fat

Figure 52. Pickleweed

Figure 53. Wishbone Bush

cells. In Desert-Holly they are particularly conspicuous and the whitish or even purplish appearance makes an attractive winter coating on this low, rounded shrub found in dry alkaline places.

In this same Goosefoot Family of small-flowered plants is Winter Fat (*Eurotia lanata*), figure 51; see also color plate 13. A low shrub with white-woolly stellate hairs, it has flowers of two types: the staminate or male consists of a 4-parted calyx and four stamens; the pistillate or female has no calyx, but a pair of bracts partly grown together and the pistil between. This plant is important for forage and is common on flats and rocky mesas mostly above 2,000 feet, from the Mojave Desert to Lassen County and eastern Washington and to Texas.

In alkaline places such as salt marshes and moist alkaline flats, the Goosefoot Family becomes fleshy and we find plants like Iodine Bush or Pickleweed (*Allenrolfea occidentalis*), figure 52. The stems are fleshy, jointed, and with minute scale-like leaves. The flowers are borne three or five in the axil of a fleshy peltate scale, in terminal fleshy spikes. Each flower has four or five lobes, one or two stamens, and a pistil. Much like it, and often also called Pickleweed, is a related genus *Salicornia*, in which the branches and flower clusters are opposite each other, while in *Allenrolfea* they are alternate.

Next we come to a different family, the Four O'Clock Family, in which the flowers are often several in a common involucre of more or less showy bracts as in our cultivated *Bougainvillea*. On the desert there grows quite commonly the Wishbone Bush or Four-O'Clock (*Mirabilis Bigelovii*), figure 53. As the name indicates, the flowers open in the late afternoon. Their white color is in the calyx, there being no petals, but one to several flowers are borne in a calyxlike involucre. The plant grows in rocky places, especially canyons, below 7,000 feet. The seedlike hard fruit is shown separately.

In the Purslane Family to which Portulaca, Bitterroot, and Miner's-Lettuce belong, the desert has SAND-CRESS (*Calyptridium monandrum*), figure 54, a small fleshy annual with a rosette of basal leaves and spreading stems having smaller reddish leaves. The flowers are tiny, with two fleshy sepals and usually three white petals. The fleshy fruit is quite elongate and flattened and projects out of the persistent calyx. Sand-Cress is common in open sandy places below 6,000 feet from Mono County south, extending too along the west side of the San Joaquin Valley and to Nevada and Sonora. It is a spring bloomer.

The Pink Family is familiar to everyone who likes the Garden Pink, Carnation, and Baby's Breath, and who appreciates weeds like Chickweed. Many of the species, although small-flowered, have character and charm and this is true on the desert. Take as an example, FROST MAT (*Achyronychia Cooperi*), figure 55, a little prostrate annual that forms small greenish-white mats on the ground by virtue of the hyaline stipules and sepals. The minute white flowers are borne in conspicuous tufts. The species is common on sandy flats and in washes, below 3,000 feet in the Colorado and eastern Mojave deserts and to Arizona. It flowers from January to May.

In the same family is DESERT SANDWORT (*Arenaria macradenia*), figure 56, a perennial from a branched woody crown which sends up slender stems one-half to one foot high. The leaves are almost needle-shaped, one to two inches

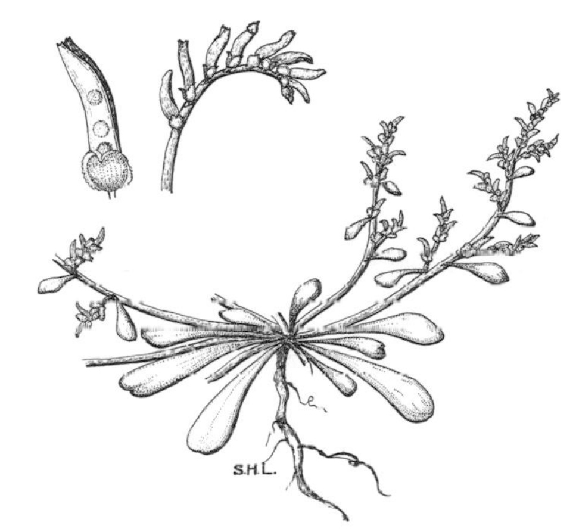

FIGURE 54. SAND-CRESS

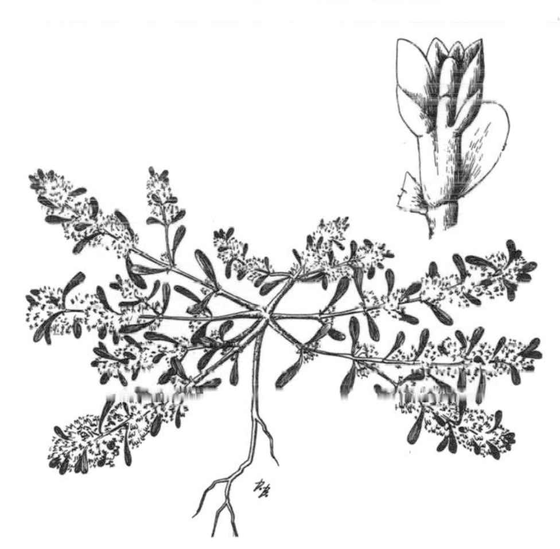

FIGURE 55. FROST MAT

FIGURE 56. DESERT SANDWORT

FIGURE 57. ROCKWORT

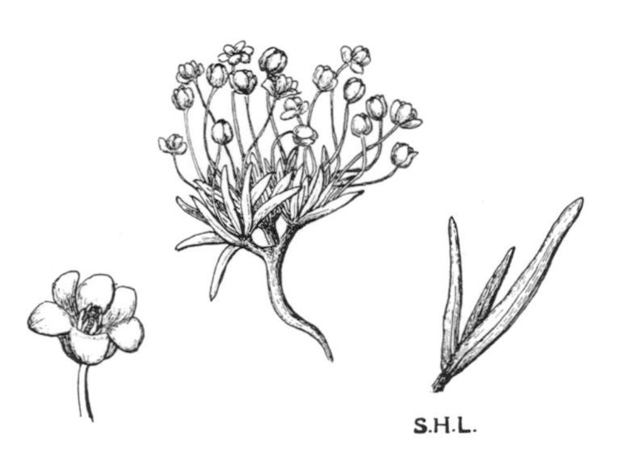

FIGURE 58. CANBYA

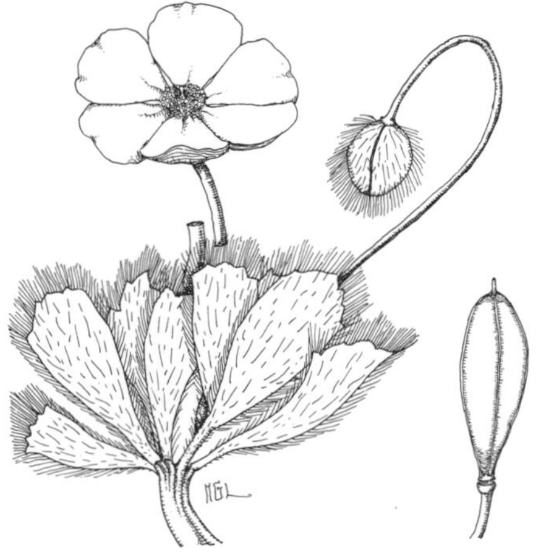

FIGURE 59. BEAR POPPY

long, borne in pairs. The stems end in open branching and bear white flowers about one-third inch across. This Sandwort is found on rocky slopes below 7,000 feet, from Riverside County north.

One of the most remarkable desert plants (still of the Pink Family) may be called ROCKWORT *(Scopulophila Rixfordii)*, figure 57. It is a low perennial arising from a dense woody root-crown, which is covered with small tufts of wool. The erect stems are two to six inches high and produce pairs of narrow leaves and small axillary clusters of tiny white flowers. The sepals are broad and with a greenish base; there are no petals but the longer outer stamens are sterile and look like narrow petals. Rockwort grows in crevices in limestone, at 3,000 to 7,500 feet in the Owens Valley and Death Valley regions, blooming from April to July.

None of our annuals is more charming than the tiny white POPPY (*Canbya candida*), figure 58. A small, tufted, almost stemless plant about an inch high, it bears narrow leaves and many small white pearly flowers, each with three sepals and six petals. It is very easy to overlook this on the sandy floor of the desert, but it is often locally common at 2,000 to 4,000 feet from the Victorville region through the western Mojave Desert to Walker Pass. Flowering is in April and May.

Another Poppy of distinction is DESERT POPPY OR BEAR POPPY *(Arctomecon Merriami)*, figure 59, a rather stout perennial, shaggy with long whitish hairs. It grows to be a foot to a foot and one-half tall, has leaves one to three inches long on equally long petioles and flowers with six white petals one to one and one-half inches long, and is thus quite a conspicuous plant. Unfortunately it is rare, being found on loose rocky slopes at 3,000 to

4,500 feet, in the Death Valley region and adjacent Nevada. To be sought for in the same region, just over the Nevada line, is the yellow-flowered *A. californica.*

When a flower is built on the plan of four and when the plant has a sharp or peppery taste, one can expect it to be in the Mustard Family, known to us by the Radish and Cabbage. The desert has many members of this family; among the white-flowered ones is the DESERT-ALYSSUM (*Lepidium Fremontii*), figure 60. A bushy rounded perennial, almost woody at the base and growing to a height of one to two feet, it bears innumerable small fragrant flowers that ultimately produce flattened pods about one-sixth of an inch long. It is common in rocky and sandy places below 5,000 feet, from northern Riverside County to Inyo County and then to Utah and Arizona. Flowering is from March to May.

FIGURE 60. DESERT-ALYSSUM

Another member of the same family is SPECTACLE-POD (*Dithyrea californica*), figure 61, an annual with coarsely toothed mostly basal leaves and, in larger plants, several more or less spreading stems. These end in elongate racemes of white flowers each over half an inch in diameter. The pods are broad and deeply notched above and below, resembling a pair of spectacles. It is a common plant of sandy places below 4,000 feet, mostly from Inyo County south, and extends into Nevada, Arizona, and Lower California. It flowers mostly from March to May.

FIGURE 61. SPECTACLE-POD

OLIGOMERIS *(O. linifolia),* figure 62, belongs to the same family as the Garden Mignonette and like that plant has the seed-pod open at the top, so that the developing seeds can be seen within. The desert plant is an odd little thing, annual, erect, a few inches to a foot tall, with linear leaves. The flowers are small, green-

FIGURE 62. OLIGOMERIS

FIGURE 63. ROCK-SPIRAEA

FIGURE 64. FERN BUSH

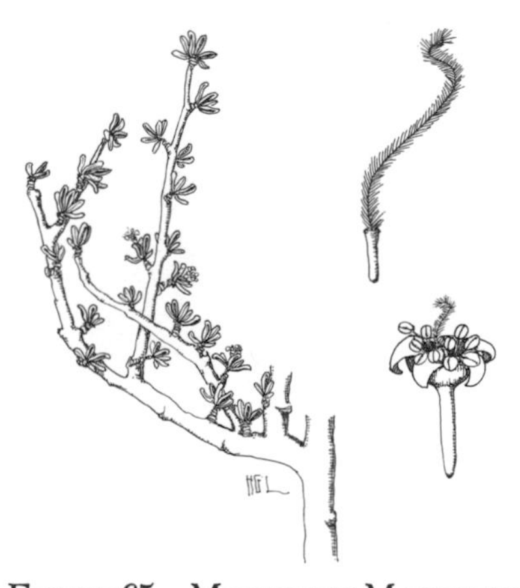

FIGURE 65. MOUNTAIN-MAHOGANY

ish, with four sepals and two petals. Oligomeris is found in open, often subsaline places below 3,000 feet, on the deserts and drier spots along the southern coast.

ROCK-SPIRAEA (*Petrophytum caespitosum*), figure 63, is a woody plant making a veritable mat on bare rocks, in the crevices of which it grows. Each flower has five sepals, five white petals, about twenty stamens, and two follicles or seed-pods—a floral structure quite typical of the Rose Family. *Petrophytum* grows on limestone ledges and rocks, at 5,000 to 9,000 feet, in the mountains of the eastern and northern Mojave Desert and east to the Rocky Mountains, from May to September.

Another member of the Rose Family is FERN BUSH or DESERT SWEET (*Chamaebatiaria millefolium*), figure 64, an aromatic shrub up to six feet high. The fern-like but rather glandular leaves and the rather large terminal clusters of flowers, each of which is about one-half inch across, are characteristic. It is found on dry rocky slopes at 3,500 to 10,200 feet, from the Panamint and Inyo-White ranges north along the east slope of the Sierra Nevada to Oregon and eastward into Wyoming. It blooms from June to August.

Still in the Rose Family, but with smaller and apetalous flowers is MOUNTAIN-MAHOGANY (*Cercocarpus intricatus*), figure 65. There are two common species on the desert, the one shown here having linear leaves less than one-half inch long and with the margins curled under almost to the midrib. The lower part of the flower is tubular and the upper deciduous bowl-shaped part with five sepals falls away after anthesis. The pistil eventually grows out from the lower tube and persists for some time as a feathery structure with a seed at the base. This species grows at 4,000 to 9,000 feet from the southern Sierra Nevada and the White Mountains to the Providence and Clark mountains. Another species, *C. ledifolius,* has broader and longer leaves.

A large group of plants in the north temperate regions is that of the Stone Fruits or *Prunus;* it belongs to the Rose Family. Among others, the desert has the DESERT ALMOND *(Prunus fasciculata),* figure 66, a much-branched deciduous shrub growing to about eight feet in height. The leaves are in bundles or fascicles; the flowers are small with petals about one-tenth inch long. The almond-shaped fruits are approximately one-third of an inch long. Desert Almond is common on dry slopes and in washes between 2,500 and 6,500 feet over much of our desert area. It flowers from March to May.

FIGURE 66. DESERT ALMOND

Quite a different *Prunus*, and one more like the familiar species in that genus, is the DESERT APRICOT *(P. Fremontii),* figure 67. Its twigs end in thorns, the leaves are broader, the flowers larger (to about one-half inch across). It is fragrant and much visited by bees. Confined to rocky places like canyons, below 4,000 feet, along the western edge of the Colorado Desert from the Palm Springs region to Lower California, the Desert Apricot begins to bloom as early as February.

FIGURE 67. DESERT APRICOT

A small family of deciduous shrubs, occurring only in southwestern North America, is the Crossosoma Family. In California we have two of the few species, one on Santa Catalina and San Clemente islands, the other, *Crossosoma Bigelovii,* figure 68, on the desert. The latter CROSSOSOMA is a stiff, much-branched, spinescent shrub, two to six feet tall, with leaves up to about one-half inch long. The white flowers are almost an inch in diameter and are followed by one to three follicles, each with two to five seeds.

FIGURE 68. CROSSOSOMA

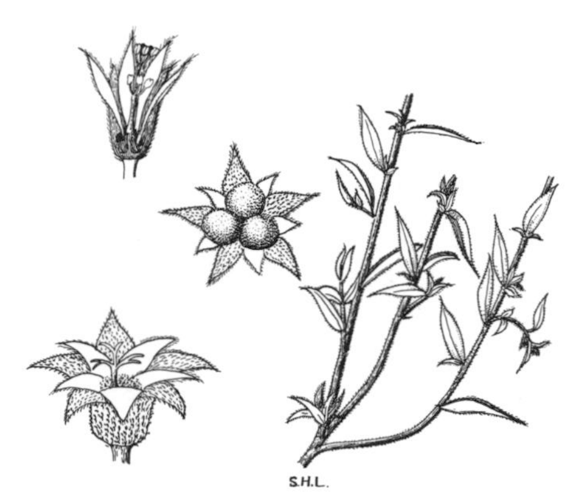

Figure 69. Ditaxis

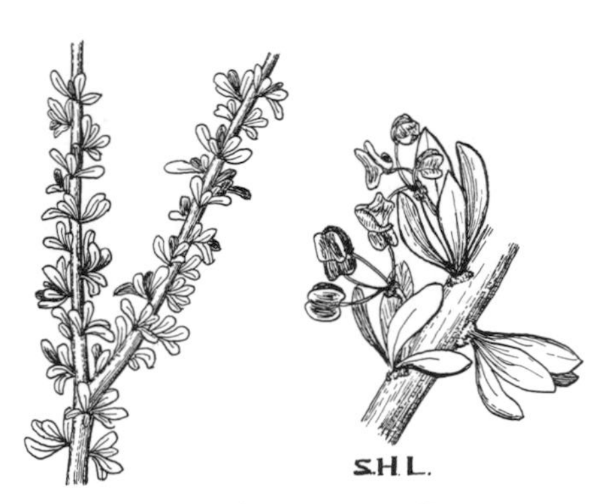

Figure 70. Purple Bush

Figure 71. Croton

Found in dry rocky canyons below 3,000 feet, it is typical of western Colorado and southern Mojave deserts. It blooms from February to April.

Among California's desert perennials is a small inconspicuous plant a few inches to a foot high, silvery-hairy on the younger twigs, and with more or less lance-shaped leaves to about an inch long. The flowers are tiny, with five sepals, five hairy petals, and a hairy three-lobed seed-pod. Ditaxis (*Ditaxis lanceolata*), figure 69, grows in rocky places and canyons below 2,000 feet in the Colorado Desert. A related species with coarse teeth at the tips of the leaves is *D. serrata* of the Mojave Desert. By their small flowers and three-lobed ovaries these two plants are characteristic of an immense family, the Spurge Family, which grows in both the Old and New Worlds and which is familiar to the gardener through a host of plants such as the Ground Spurge, the Poinsettia, and the Castor-Bean.

Another member of the family is Purple Bush (*Tetracoccus Hallii*), figure 70, of the region between the Mojave and Colorado deserts, ranging from the Eagle, Cottonwood, and Chuckawalla mountains to near Needles and Ivanpah and into southwestern Arizona. It is an erect shrub, about two to six feet high, with stiff branches bearing small clusters or fascicles of small leaves on short spurlike branchlets. Male and female flowers are borne on separate plants, each having four to six sepals but no petals. Leaves are less than half an inch long.

To this family also belongs Croton, which in the tropics contains a number of shrubs with bright-colored foliage and is often grown as an ornamental. Here our desert offers a very modest species, a form of *Croton californicus*, figure 71. It is a perennial herb, mostly one to two feet high, with a number of branching stems, which together with the leaves are hoary or pale olive-green. The leaves are scarcely an inch long; the flowers are very small, without petals. Croton is found in sandy places with Creosote Bush.

As additional examples of the Spurge Family and with the three-lobed ovary are two members of the genus *Stillingia*. One of them (*S. linearifolia*), figure 72, is a strong-rooted perennial, a foot to two feet tall, with slender loosely branched stems and linear leaves. Staminate and pistillate flowers are separate but borne on the same plant, both being shown in the illustration. This plant is occasional in washes and rocky places below 3,500 feet, in deserts and sometimes coastal areas. It flowers from March to May.

Our other STILLINGIA here shown (*S. spinulosa*), figure 73, is an annual, less than a foot high, with spreading branches and broad, toothed, three-nerved leaves. The flowers are borne in spikes, male above and female at the base. This nice green little plant is frequent in dry sandy places below 3,000 feet, in the Colorado and Mojave deserts and into adjacent Nevada and Arizona. The flowers can be seen between March and May.

As recognized by more conservative botanists, the genus *Euphorbia* has perhaps a thousand species. Among them is our cultivated Poinsettia and in the same part of the genus is a desert annual, erect, green, one-half to one and one-half feet tall, freely branched and bearing narrow leaves one to two inches long. It is DESERT POINSETTIA (*Euphorbia eriantha*), figure 74. The upper leaves form whorls around the flower clusters as do the large red leaves of the cultivated Poinsettia. The desert plant occurs in rocky places, like

FIGURE 72. STILLINGIA

FIGURE 73. STILLINGIA

FIGURE 74. DESERT POINSETTIA

FIGURE 75. GOATNUT

FIGURE 76. DESERT-JUJUBE

FIGURE 77. SANDPAPER PLANT

canyons and mesas, below 3,000 feet, from the Eagle Mountains and Andreas Canyon (near Palm Springs) to Mexico and Texas. It flowers from March to April.

GOATNUT or JOJOBA (*Simmondsia chinensis*), figure 75, is the only native California plant belonging to the same family with cultivated Box, the hedge plant. It is a stiff-branched shrub three to six feet tall, with ascending leaves and small greenish flowers. Male and female are borne on separate plants; see the drawings to the right and upper left respectively. The three-angled nut, almost an inch long, is oily and was used as food by the Indians. It is common on dry barren slopes below 5,000 feet, from Little San Bernardino Mountains to Imperial County and west to inland coastal valleys. It blooms from March to May.

The name DESERT-JUJUBE is applied to a desert shrub (*Condalia Parryi*), figure 76. It is five to fifteen feet tall, with spiny flexuous branchlets and broad leaves to about two-thirds of an inch long. The flowers are minute, but the conspicuous feature is the plumlike fruit that hangs on until late summer and turns yellow. This Jujube is local on dry slopes and in canyons along the western edge of the Colorado Desert from Morongo Pass to Lower California. It blooms early in the spring and should not be confused with the similar but larger Desert Apricot of the same region (see p. 35).

SANDPAPER PLANT is a fitting name for *Petalonyx Thurberi*, figure 77, since its surface is covered with short barbed hairs that feel rough to the touch. Woody at the base, with many low, spreading grayish stems one to two feet long, it has numerous sessile leaves one-fourth to one inch long and terminal clusters of fragrant white flowers with five sepals and five petals. Its barbed hairs make its leaves adhere to suede jackets, woolly sweaters, and socks, as many a walker has learned. It is much visited by bees and is sometimes called Honeybush.

Now we come to a flower built on the plan of four, a member of the Evening-Primrose Family and resembling the Mustard Family in this regard, but differing in having the ovary (seed-bearing part) beneath the flower instead of up in it. GAURA (our species, *G. coccinea*), figure 78, is a bushy perennial with leaves to about an inch long and white flowers one-half inch in diameter, which turn pinkish in age. The capsule is hard and nutlike. The species grows particularly about limestone in the mountains of the eastern Mojave. It blooms from April to June.

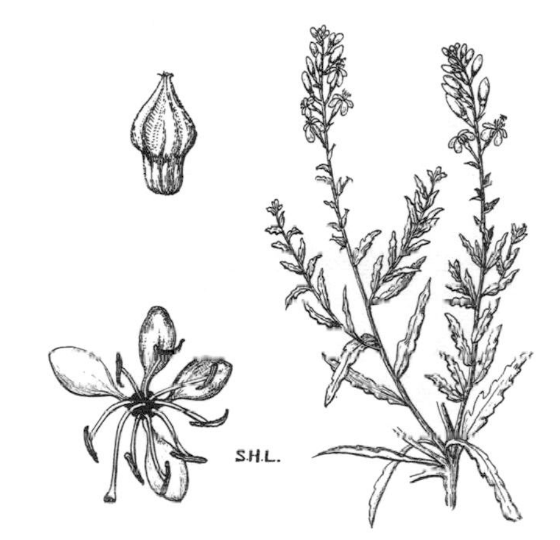

FIGURE 78. GAURA

A spiny little shrub of the Olive Family, which means that its flower is apt to be constructed on the plan of four, is TWINFRUIT *(Menodora spinescens)*, figure 79. Not more than a couple of feet high, spreading and with irregularly divergent spiny branches, it bears small leaves that may get one-half inch long. It has five to seven narrow sepals about one-sixth of an inch long, white petals with brownish-purple backs, and two-lobed capsules. It is found on dry mesas and slopes, mostly between 3,500 and 6,500 feet, from Owens Valley to the eastern Mojave Desert and into Nevada and Arizona.

FIGURE 79. TWINFRUIT

A genus primarily of North America and Japan is AMSONIA, the species shown being *A. brevifolia*, figure 80. Amsonia with us is a perennial herb, a foot or more tall, with slightly bluish flowers that soon fade whitish. The pods are two to three or more inches long, much like those of Indian-Hemp and Oleander to which it is related. In addition to this hairless green-leaved species and growing with it

FIGURE 80. AMSONIA

Figure 81. Milkweed

Figure 82. Milkweed

Figure 83. Fiesta Flower

in dry places between 2,500 and 6,000 feet, is another, *Amsonia tomentosa*, which is white-woolly. Both species are found from the northern edge of the Colorado Desert to the Panamint Mountains; they bloom from March to May.

With milky sap as in Amsonia is the true Milkweed Family, represented here by Milkweed (*Asclepias erosa*), figure 81. The species is perennial, about two feet tall, and has pairs of broad leaves up to six inches long. The round clusters of many flowers and the typical milkweed capsules with many flat seeds, each of which bears a tuft of white hairs, are typical. It is well distributed on the deserts below 5,000 feet and extends into the San Joaquin Valley and Utah and Arizona. It blooms between May and July.

Another quite different Milkweed (*Asclepias subulata*) is represented in figure 82. It is almost shrubby with its many rushlike almost leafless stems three to five feet high. Leaves when present are filiform, one to two inches long. The greenish-white flowers are about one-fourth inch long. Occasional in desert washes and sandy places below 2,000 feet, it is found in the Colorado Desert and eastern Mojave Desert. Another similar species, in that the leaves drop off soon, is *A. albicans*, having white-waxy stems and growing in rockier places.

Californians who know the out-of-doors certainly are acquainted with Baby-Blue-Eyes or Nemophila and with Wild-Heliotrope (see page 46). In the same family is an inconspicuous but often common desert annual with weak sprawling stems and smallish white flowers, sometimes called Fiesta Flower (*Pholistoma membranaceum*), figure 83. Each corolla lobe has a small purplish narrow spot. Usually found in shaded places, as under a bush or overhanging rocks, it grows below 3,500 feet from Inyo County to Imperial County, then west along the inner Coast Ranges to Contra Costa County. It blooms from March to May. The drawing shows the peculiar calyx and seed.

In the true Heliotrope or Borage Family, in which the flowers are mostly in coiled cymes and the plant often has spiny prickly hairs, is CRYPTANTHA, sometimes called WHITE FORGET-ME-NOT. The species shown in figure 84 is *Cryptantha nevadensis*. It is a slender-stemmed annual to about a foot and one-half high, with appressed hairs, and very small almost tubular flowers. The calyx is bristly. There are many similar species on the desert, some with larger flowers, some of lower stature, some annual, some perennial, some with flowers yellowish. They mostly grow in well-drained sandy or gravelly places and are spring bloomers.

In this same Heliotrope Family is another group, PECTOCARYA, figure 85, often conspicuous in California deserts and coastal areas, not because of its size but because it grows in such large masses. Only a few inches high, usually much branched, with tiny white flowers and with four divergent one-seeded nutlets in place of a seed-pod, it often covers large sandy areas in between bushes. There are several species that differ in technical characters. It blooms in spring.

The Nightshade Family, with Potato, Tomato, Eggplant, Bell-Pepper, Tobacco, and Petunia, is of great importance to man. Relatively insignificant, then, appears BOXTHORN, a group of armed shrubs, of which several are found in the California deserts. One of these, *Lycium Cooperi*, figure 86, is densely leafy, three to six feet high, with leaves to about one inch long, and with greenish-white tubu-

FIGURE 84. WHITE FORGET-ME-NOT

FIGURE 85. PECTOCARYA

FIGURE 86. BOXTHORN

FIGURE 87. BOXTHORN

FIGURE 88. DESERT TOBACCO

FIGURE 89. JIMSONWEED

lar flowers almost half an inch long. The dry greenish fruit is constricted toward its summit. This Boxthorn is found in dry places below 5,000 feet in both our deserts, in the upper San Joaquin Valley, and to Utah and Arizona. Flowers come from March to May.

Another BOXTHORN or DESERT THORN is *Lycium brevipes,* figure 87. The flower is white or nearly so and is about one-third of an inch in length. This very divaricately branched shrub occurs in washes and on hillsides below 1,500 feet, along the western edge of the Colorado Desert to Lower California and to Sonora. Flowers come in March and April. The fruit is a small bright red round berry. Several other species in our deserts vary in leaf shape, flower size, and color.

DESERT TOBACCO *(Nicotiana trigonophylla)*, figure 88, is a viscid mostly perennial herb found about rocky places. It is one to two and one-half feet tall, with fairly large leaves clasping at the base. The greenish-white flowers are almost an inch long and the dried plant has been used by Indians for smoking. This plant occurs at below 4,000 feet from Mono County south to Mexico and east to Texas. Flowering season is mostly from March to June. As in our other California wild tobaccos, the foliage is heavy-scented.

In the same Nightshade Family and with the characteristic rank odor of the leaves is JIMSONWEED (*Datura meteloides*), figure 89, a perennial to about two feet high, coarse-leaved, and with large trumpet-shaped flowers four to eight inches in diameter. They are white, suffused with violet. The species grows in open places below 4,000 feet over most of southern and central California and east to Texas, blooming from April to October. Its smaller-flowered relative (*D. discolor*) of the Colorado Desert was used by the Coahuila Indians in initiation to manhood.

Our next plant is a small annual PLANTAIN (*Plantago insularis*), figure 90, and much more delicate and less coarse than the big introduced plantains that occur so commonly as weeds in lawns and waste places. We have in California a number of these small annuals, this species being characterized by its reddish-yellow shining seeds. The flower has four persistent papery petals and the seeds are flattened. The distribution is in open sandy places below 4,500 feet, from the deserts to the coast and east to Utah and Arizona. The flowering season is from January to April.

In the Gourd Family I always expect melonlike plants with fairly large and often many-seeded fruits. But, as so often happens, the desert comes up with a surprise in BRANDEGEA (*B. Bigelovii*), figure 91. The vine resembles that of the Wild-Cucumber, familiar to many who grew up in the Middle West, but the fruit is a small one-seeded flattened structure less than half an inch long. The flowers are small, the male or staminate being barely one-twelfth of an inch across. It is local in washes and canyons below 2,500 feet from the southern Mojave Desert to Lower California and Arizona.

The Sunflower Family is an unusually large one in the desert; it can always be recognized by having numerous small florets wrapped up in an outer involucre so that this whole head simulates a solitary flower, often having the outer florets modified into petallike ray-flowers. Such a plant is the little annual, the DESERT STAR (*Monoptilon bellioides*), figure 92. Growing flat on the desert floor, it is stiff-hairy, from an inch or so to several inches in diameter. It is a typical "daisy"

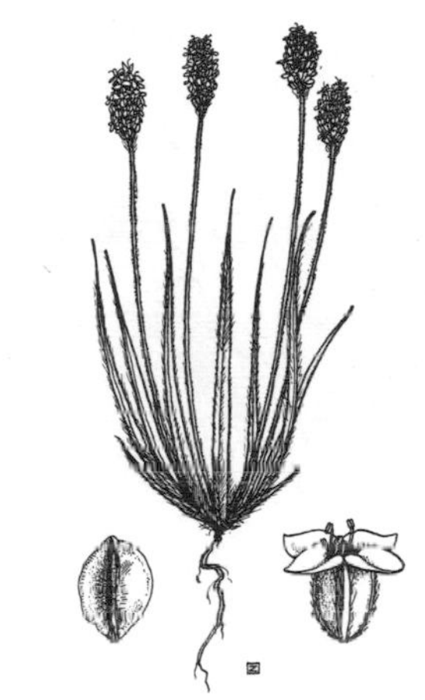

FIGURE 90. PLANTAIN

FIGURE 91. BRANDEGEA

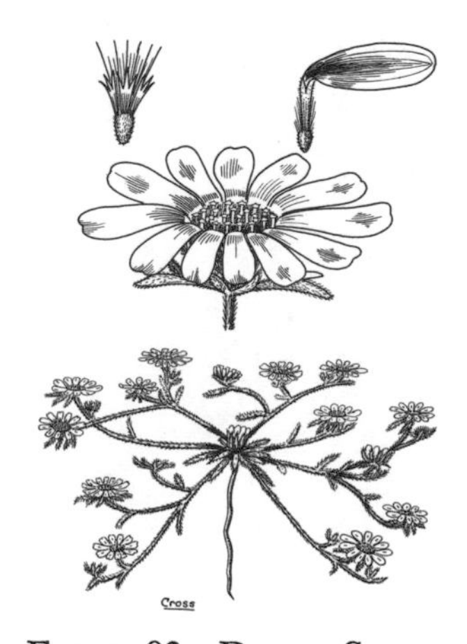

FIGURE 92. DESERT STAR

FIGURE 93. MULE FAT

FIGURE 94. BUGSEED

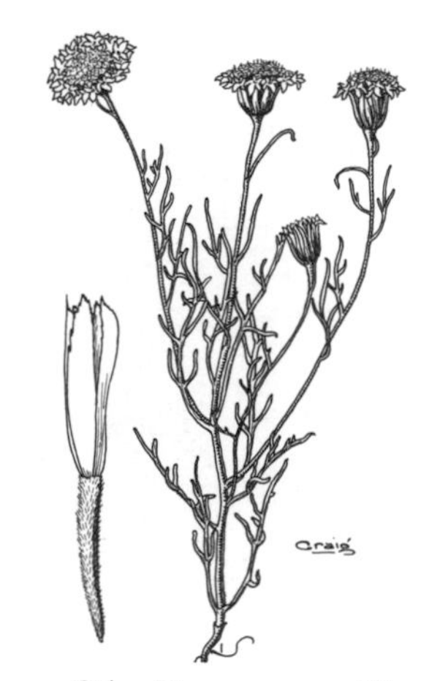

FIGURE 95. PINCUSHION FLOWER

with white or pinkish ray-flowers and is abundant on sandy or stony desert plains below 3,000 feet.

A willowlike shrub six to twelve feet tall, MULE FAT (*Baccharis viminea*), figure 93, has mostly almost entire leaves (that is, not toothed or lobed), which are somewhat glutinous. The small flower heads are borne in terminal clusters, each floret producing a seedlike fruit with a tuft of hairs at the tip. Mule Fat grows mostly along washes and ditch banks below 1,500 feet, is common on coastal slopes, but extends across the desert to Utah and Arizona. Flowers can be found at most seasons.

BUGSEED OR DICORIA (*D. canescens*), figure 94, is most conspicuous in midwinter in sandy places along roadsides and open places. It is an annual, one to three feet high, whitish or grayish and coarsely haired, with entire or toothed leaves. The heads are many and small at the time of flowering, the male remaining so, but the female having one or two flowers subtended by bracts that enlarge and become thin; both types are shown at the upper right. In a variety of forms the plants are found on both Mojave and Colorado deserts and to the east of California.

PINCUSHION FLOWER (*Chaenactis Fremontii*), figure 95, is a good name for a green erect annual which may attain a foot in height and has rather fleshy leaves often with linear lobes. The heads are pincushionlike, being tubular like the inner but somewhat enlarged. Each floret produces a seedlike fruit with terminal papery scales. This species is common on sandy mesas and open slopes below 3,500 feet on both deserts and runs along the edges of the San Joaquin Valley to Kettleman Hills. It flowers from March to May.

Other species have more finely divided leaves and some have yellow flowers.

Some groups in the Sunflower Family have all the florets in a head strap- or petal-shaped, as in the common Dandelion. A striking example is TOBACCO-WEED *(Atrichoseris platyphylla),* figure 96, a smooth annual with a rosette of fleshy, sometimes spotted leaves flat on the ground and an erect leafless stem one to several feet tall. At the summit are spreading heads of pure white flowers that are very pleasantly fragrant, and appear from March to May. It is common in sandy washes in the more eastern parts of both deserts as far north as Death Valley and ranges beyond our borders to Utah and Arizona.

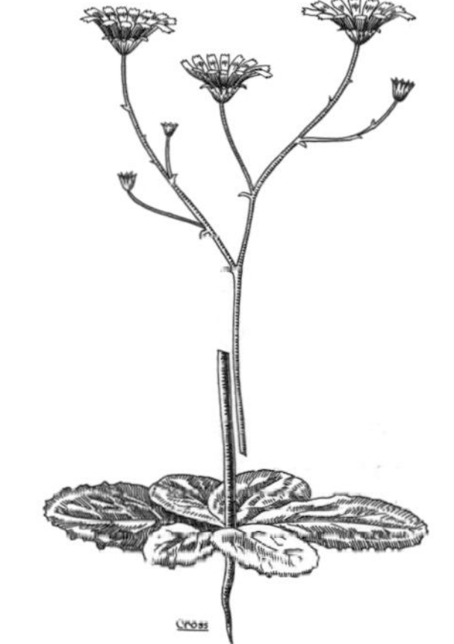

FIGURE 96. TOBACCO-WEED

DESERT-CHICORY *(Rafinesquia neomexicana),* figure 97, is a rather weak-stemmed annual with white ray-flowers that are veined rose-purple on the outer side. The heads are rather large and very fragrant. A closely related species is *R. californica* found on coastal slopes as well as desert, but *R. neomexicana* with larger flowers is confined to the desert, where it is common in the shade of shrubs and in canyons, its range extending to Utah and Texas. It blooms from February to May.

FIGURE 97. DESERT-CHICORY

FLOWERS BLUE TO VIOLET

Section Four

In the Buttercup Family (see page 15) are Buttercup, Larkspur, Anemone, Columbine, and many other familiar plants. The family is less common on the desert than in more moist places, but a LARKSPUR (*Delphinium Parishii*), figure 98, occurs in various color forms, exhibiting a typical light blue from the Mojave Desert to Coachella Valley, a deeper blue from the Santa Rosa Mountains to Lower California, and a pinkish purple in the Mount Pinos region. It ascends slopes to 7,500 feet and flowers from March to June.

In the Pea Family is a number of more or less woody plants of the genus *Dalea* having stems, leaves, and calyces dotted with glands. One of these protected by law is SMOKE BUSH or SMOKE TREE (*Dalea spinosa*), figure 99, so called because it consists of an intricate mass of slender thorny branches, from which the leaves soon drop and its ashy-gray pubescence makes it look from a distance like puffs of smoke. It attains a height of three to twenty feet. The bright blue-purple pea-shaped flowers come in June and July after the leaves fall off. It occurs in sandy washes below 1,500 feet, from Daggett east and south, ranging to Arizona and Sonora.

Another Dalea is the type known as INDIGO BUSH (*Dalea Fremontii*), figure 100. An intricately branched shrub to about six feet tall, it has grayish younger growth and dark purple-blue flowers. It varies in hairiness and shape of leaflets and is distributed on dry slopes below 5,000 feet, from Owens and Death valleys to Riverside County and east. The flowers come in the spring. Like other species it has conspicuous glands, especially on the calyx. See also page 37.

FIGURE 98. LARKSPUR

FIGURE 99. SMOKE BUSH

FIGURE 100. INDIGO BUSH

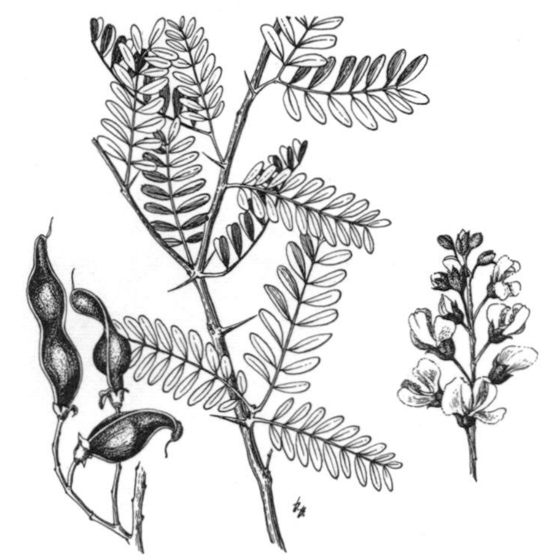

FIGURE 101. DESERT-IRONWOOD

FIGURE 102. PSORALEA

FIGURE 103. GILIA

DESERT-IRONWOOD *(Olneya Tesota)*, figure 101, is another member of the Pea Family and one of the larger plants on our desert, forming spinose trees with broad grayish crowns and from fifteen to twenty-five feet high. The leaves are compound, the flowers are violet-purple, almost half an inch long, and come in profusion when the tree blooms. The species is found in desert washes and on sandy fans below 2,000 feet on the Colorado Desert. It ranges to Arizona, Sonora, and Lower California, flowering late in the spring. It was named for the hardness of its wood. The parched seeds were eaten by the Indians.

A group close to Dalea in its glandular foliage is PSORALEA (*Psoralea castorea*), figure 102. This desert species is an almost stemless perennial with appressed hair, trifoliolate leaves, and headlike clusters of blue flowers almost half an inch long. It is found in sandy flats and washes at 1,500 to 3,000 feet, on the Mojave Desert from Victorville to Yermo and on to Utah and Arizona. The flowers appear in April and May. The starchy taproots were eaten by the Indians.

GILIA of the Phlox Family (see pages 21, 44, 100) is here represented by *Gilia latiflora*, figure 103, a variable group of erect annuals with leaves largely near the base of the plant and stems to about one foot high. The flowers may vary from half an inch to one inch long and from blue to violet, with a throat yellow to purple. Occurring in great masses in sandy flats between 2,500 and 5,000 feet, they add greatly to the color on the deserts in a good year, being found from inner San Luis Obispo and Santa Barbara counties to the western Mojave Desert. Flowering is from March to May.

LANGLOSIA (see pages 21, 45) is a

relative of Gilia and represented here by *L. setosissima*, figure 104, a tufted annual a couple of inches high or with short prostrate branches. The leaves are less than an inch long and bear teeth especially toward the tip. The flowers are light violet, not spotted, about one-half inch long. Distribution is in dry sandy places below 3,500 feet and east to Idaho, Utah, and Sonora. Flowering is from April to June. A related species, agreeing with this one in having the corolla lobes all much alike, but with purple dots, is *L. punctata* (see page 45).

In the same family is ERIASTRUM, with linear leaves or leaves divided into linear segments and mostly bristle-tipped. A common desert annual is *E. eremicum*, figure 105, erect or spreading, a few inches to almost one foot high, more or less woolly, with three to twenty flowers in each woolly cluster. The corolla is over one-half inch long, two-lipped, violet in color. The species is found in sandy places below 5,000 feet, on both deserts, ranging from Inyo County to Imperial County and blooming from April to June.

WILD-HELIOTROPE (*Phacelia distans*), figure 106, is in the large genus *Phacelia* (see pages 46, 47) and is an annual, one to three feet high. Frequently it is branched and even somewhat spreading. The fernlike leaves and coiled cymes of smallish blue or bluish flowers are characteristic. It is common in fields, slopes, and canyons in the deserts and in coastal California as far north as Mendocino County. Eastward it ranges to Nevada and Sonora, flowering from March to June.

Another PHACELIA is YELLOW THROATS (*P. Fremontii*), figure 107, an annual, with spreading stems to almost one foot long. It is quite glandular in its upper parts

FIGURE 104. LANGLOISIA

FIGURE 105. ERIASTRUM

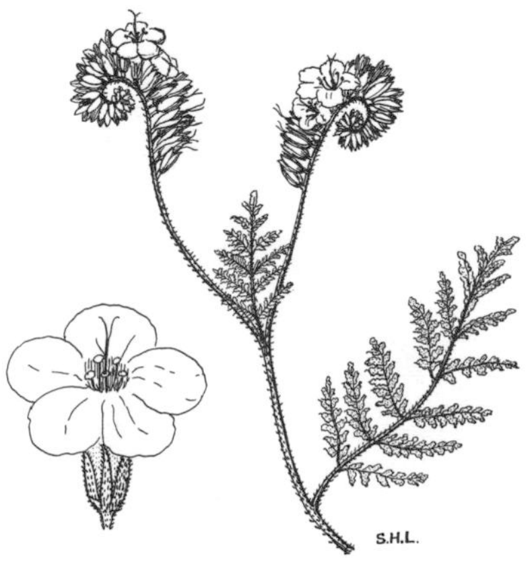

FIGURE 106. WILD-HELIOTROPE

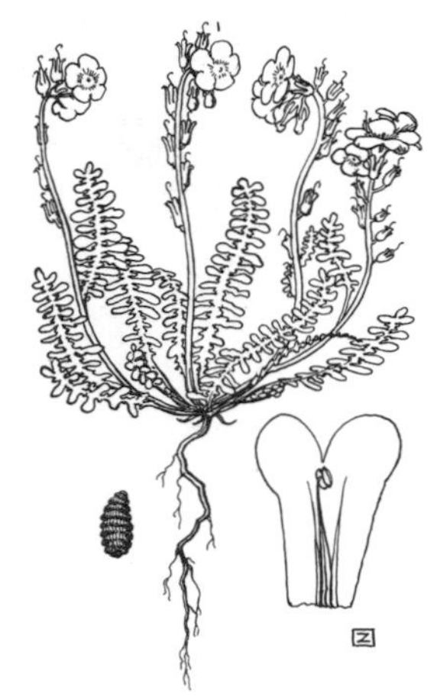

FIGURE 107. YELLOW THROATS

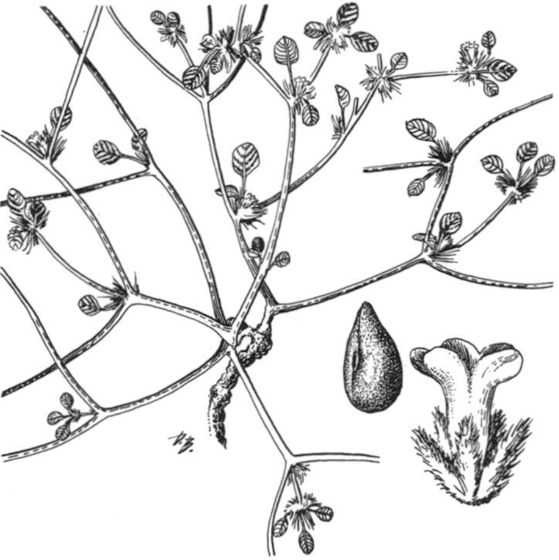

FIGURE 108. COLDENIA

FIGURE 109. MOHAVE SAGE

and short-hairy below. The bright blue to deep lavender flowers are almost half an inch long and have a yellow tube. From sandy or clayey slopes or flats below 7,000 feet, it ranges from Inyo County to Riverside County and the Santa Rosa Mountains and along the western edge of the San Joaquin Valley, as well as eastward into Utah and Arizona. Flowers are from March to May.

In the related Borage Family (see page 75) in which the fruit is apt to consist of small hard nutlets, is COLDENIA (*C. plicata*), figure 108. This species is a matted perennial from a deep woody root, with prostrate stems to about one foot long. The leaves have four to seven pairs of deeply impressed veins and are rather grayish-hairy with closely appressed hairs. The small flowers (one-sixth inch long) are almost hidden. Growing in sandy places below 3,000 feet, it is found in the Colorado Desert and the eastern Mojave Desert, and in Nevada and Arizona. It flowers between April and July. Another similar species, *C. Palmeri*, has only two or three pairs of veins.

When I first came to California I was impressed by the very different species of Sage or Salvia, of the Mint Family (see pages 48, 87). One of these, the aromatic MOHAVE SAGE (*Salvia mohavensis*), figure 109, is a compact many-branched shrub from one to more than two feet tall, with leaves almost an inch long and having deeply furrowed veins. The flowers are pale blue or lavender and over one inch long. It grows in dry rocky washes and canyons at 1,000 to 5,000 feet, from the Little San Bernardino and Sheephole mountains to the Clark and Turtle mountains and into Nevada and Sonora. The flowers are from April to June. Another somewhat similar species, *S. eremostach-*

ya, has larger flowers subtended by purplish-green instead of pale bracts. It grows south of the Santa Rosa Mountains.

In the Mint Family is also DESERT-LAVENDER (*Hyptis Emoryi*), figure 110, an erect aromatic shrub three to ten feet high with whitish somewhat closely-woolly leaves. The flowers are violet, two-lipped, about one-sixth of an inch long. Common in washes and canyons below 3,000 feet. Desert-Lavender is found in the Colorado Desert and southern Mojave Desert and to Arizona and Sonora. It blooms from January to May and is much visited by bees.

FIGURE 110. DESERT-LAVENDER

BLADDER-SAGE (*Salazaria mexicana*), figure 111, is of the Mint Family also and is aromatic, has leaves in pairs (opposite), two-lipped corollas, and nutlets instead of seed-pods. It is a low intricately branched shrub with grayish lance-shaped leaves about one-half of an inch long. The purplish-blue corolla of about the same length has dark lips and pale throat. See also page 48. This species is common in dry washes and canyons below 5,000 feet, from Inyo County to Riverside County and east to Utah and Texas. The flowers, which appear from March to June, are followed by greatly inflated bladdery calyces.

FIGURE 111. BLADDER-SAGE

DESERT-ASTER (*Machaeranthera tortifolia*), figure 112, is to my mind one of the most attractive of the desert plants. A strong perennial, somewhat woody at the base, it sends up a number of branches one to two feet high with elongate rather sharply toothed leaves and long-stemmed flower heads displaying yellow centers within blue-violet to lavender to even pinkish ray-flowers. These heads may be two inches in diameter. Desert-Aster is found in dry rocky places between 2,000 and 5,500 feet, from the northern Colo-

FIGURE 112. DESERT-ASTER

FIGURE 113. DESERT FLEABANE

rado Desert to the White Mountains of Inyo County and eastward to Utah and Arizona. It blooms mostly from March to May, but sometimes also in the autumn.

DESERT FLEABANE (*Erigeron foliosus* var. *Covillei*), figure 113, is a perennial with foliage that feels harsh to the touch. The leaves are narrow and the heads of blue flowers are about one inch across. It is a typical "daisy," with central yellow tubular florets and marginal blue narrow ray-florets that are petallike. The plant multiplies by underground shoots as well as by seed, so that it often forms patches several feet across. It is found on grassy or brushy slopes below about 6,000 feet, from the north base of the San Bernardino Mountains to Inyo County and on both the east and, more sparingly, the west flanks of the Sierra Nevada. It flowers from May to August. It commemorates the name of Frederick V. Coville who botanized in Death Valley in 1891.

FLOWERS YELLOW TO ORANGE

Section Five

For those of us who grew up in a cooler climate than California, it is difficult to become accustomed to plants combining lilylike flowers and a woody base. But so it happens in our dry warm southwestern states, where yuccas become trees and century plants send up such woody stems that they can be used for ridge poles in houses. The most northern, hence most modest, of such plants is a CENTURY PLANT (*Agave utahensis*), figure 114, of the eastern Mojave Desert. With a tuft of fleshy spinose grayish leaves up to almost a foot long, the plant sends up after some years a slender flowerstalk about five to seven feet high. The yellow flowers are an inch or more long, with outer and inner rows of three petallike segments.

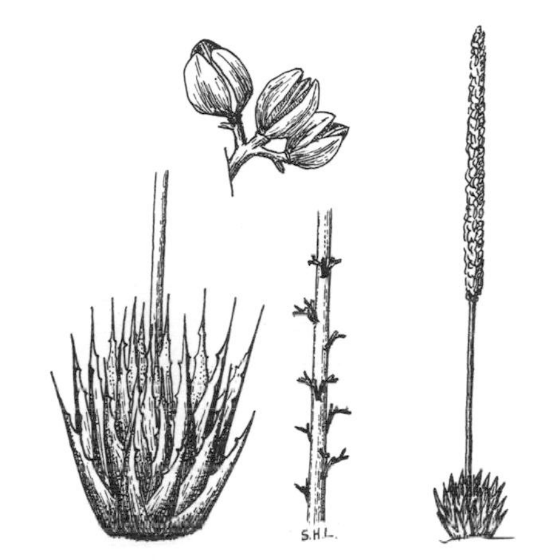

FIGURE 114. CENTURY PLANT

The Wild-Buckwheat (see page 30) is so large a group in California that the small flowers range widely in color. In figure 115 is shown the DESERT TRUMPET (*Eriogonum inflatum*), a perennial with basal leaves, inflated stems to over two feet tall, and very slender ultimate branches bearing minute hairy yellow flowers with three outer and three inner petallike segments. The plant is common in gravelly and rocky places below 6,000 feet, from Mono County south and east to Lower California, Arizona, and Utah. Flowers come in the spring and fall.

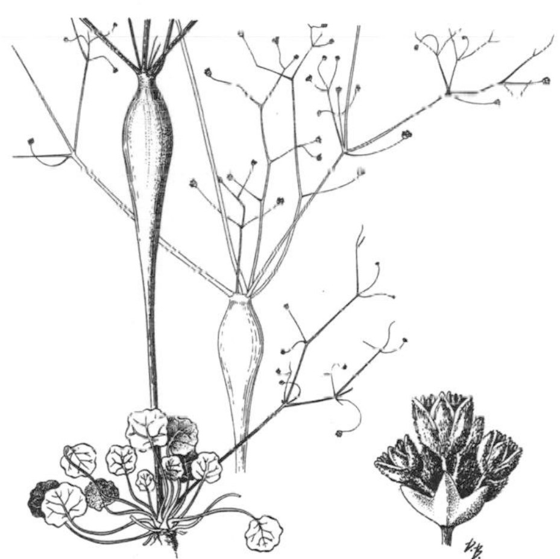

FIGURE 115. DESERT TRUMPET

Related to the Buckwheat but with the flowers enclosed in an involucre ending in spinelike or bristlelike teeth is CHORIZANTHE. A typical desert species is *Chorizanthe brevicornu,* figure 116, a small yellowish-green, brittle-stemmed annual with narrow basal leaves and minute flowers. The involucre surrounding these has recurved teeth; the actual flower is whitish, but is so small and the general effect of the plant is so yellowish that the species is placed in this section. It is common

FIGURE 116. CHORIZANTHE

FIGURE 117. HONEY SWEET

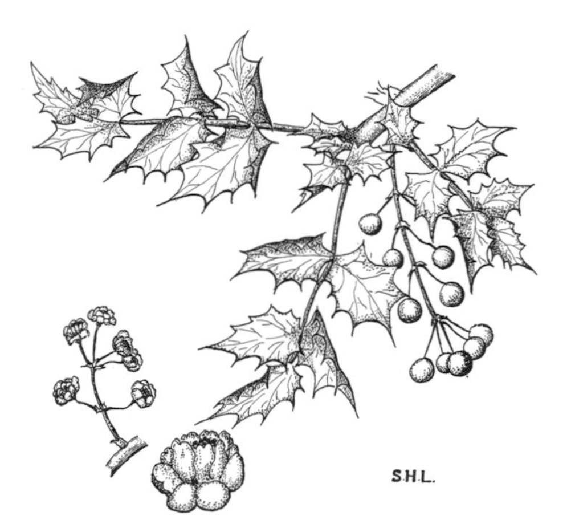

FIGURE 118. BARBERRY

FIGURE 119. CREAM CUPS

in dry stony and gravelly places below 5,000 feet, over most of our California deserts, blooming from March to June.

A white woolly perennial with a low broad habit and growing to about one foot high and twice as wide, is HONEY SWEET (*Tidestromia oblongifolia*), figure 117. It is related to Amaranth (see page 13) and has very small yellowish flowers with a pleasant odor. They are borne in small clusters surrounded by bracts, which may turn reddish late in the season. This plant is found in dry sandy places, such as washes, generally below 2,000 feet altitude, through the Colorado and eastern Mojave deserts to Death Valley and then into Nevada and Arizona. It blooms from April to November.

The BARBERRY is another plant not seemingly adapted to the desert, but two or three species, with their spiny harsh leaves protected from browsing and adapted to dry conditions, do occur in the desert mountains. One is *Berberis Fremontii*, figure 118, with stiff stems three to eight feet tall. The flowers have circle after circle of petallike parts; the berries are yellow to red and become dry and inflated. Found at 3,000 to 5,000 feet in the eastern Mojave Desert and west to Cushenbury Springs, it is perhaps not as common as *B. haematocarpa* (see page 32), which has purplish-red juicy berries. It is found in the New York and Old Dad-Granite mountains and eastward.

For one of the Poppy Family, see page 32. Another member is CREAM CUPS (*Platystemon californicus* var. *crinitus*), figure 119, a soft hairy annual growing to almost one foot tall, with nodding buds and yellowish petals one-third to one-half inch long. It is locally common in sandy places from Cuyama Valley of Santa Barbara County in the South Coast Ranges to the deserts along their western edge to Lower California. It is a spring bloomer.

In the Mustard Family (see pages 33 and 67) are many plants whose stem and leaves or whose seeds have a sharp taste. Such a plant is TANSY-MUSTARD (*Descurainia pinnata*), figure 120, an erect, more or less branched annual with finely dissected leaves. The small rather pale yellow flowers are numerous, each forming an elongate seed-pod with many small yellowish to brown seeds. This species has several named variants. It is common on the deserts and far beyond, in gravelly and sandy places. Not many years ago an Indian basket was found stowed away under an overhanging rock near Twentynine Palms with some quarts of seeds, having been there apparently for many years. Flowering is in the spring months.

FIGURE 120. TANSY-MUSTARD

WALLFLOWER (*Erysimum capitatum*), figure 121, is another member of the Mustard Family, with erect simple or branched stems to about two feet tall, with very fragrant yellow to orange flowers, and with erect elongate pods. It occurs in dry stony places below 8,000 feet on the Mojave and on coastal slopes, ranging north to British Columbia and Idaho. It blooms from March to July.

FIGURE. 121. WALLFLOWER

BLADDER POD or BEAD POD is *Lesquerella Palmeri*, figure 122, an annual, with slender spreading or ascending stems to about one foot long and with bright yellow petals one-fourth inch long. The fruits are almost round. It grows in sandy places below 3,500 feet, from the eastern Mojave Desert and northeastern Colorado Desert to Utah and Arizona. Flowers appear from March to May. Another species is *Lesquerella Kingii*, a silvery-coated perennial on dry rocky slopes at 5,000 to 9,000 feet in mountains of the Mojave.

FIGURE 122. BEAD POD

One of California's early botanical ex-

FIGURE 123. LYRE POD

FIGURE 124. STINKWEED

FIGURE 125. OXYSTYLIS

plorers from the Old World was Thomas Coulter, who crossed the Colorado Desert from Pala in San Diego County on his way to Mexico. In the rocky canyons along the western edge of the desert he found a plant later named for him and now sometimes called LYRE POD (*Lyrocarpa Coulteri*), figure 123. Related to Spectacle Pod (page 67), it has larger fruits and tawny narrow petals. It is perennial, and blooms from December to April.

Close to the Mustard Family and also with four petals, but usually with very ill-smelling foliage, is the Caper Family. STINKWEED (*Cleomella obtusifolia*), figure 124, is a diffusely branched annual, often with long, rather trailing stems. The petals are almost one-fourth inch long, the two-lobed capsule about as broad. This plant grows in alkaline flats below 4,000 feet, from Inyo County to the Colorado Desert and into Nevada and Arizona. Its season is a long one, flowers continuing to appear from April to October.

Also in the Caper Family is *Oxystylis lutea*, figure 125, an annual having yellow stems up to three feet high. The flowers are crowded in headlike clusters and are scarcely one-twelfth of an inch long. Each flower produces two one-seeded nutlets. OXYSTYLIS too inhabits alkaline flats and washes and is found at elevations below 2,000 feet, from the Death Valley region to Tecopa and adjacent Nevada, flowering from March to October. The dried stems with the persistent spiny remains of the styles (see figure A in the Introduction) are often conspicuous long after the death of the plant.

The Rose Family (see pages 68, 69) has in the genus *Potentilla* a large number of species from mountain meadows and similar places. It is interesting, then, to find one growing in rock-crevices on the desert. This POTENTILLA (*P. saxosa*), figure 126, is a small perennial with thick woody root and slender glandular stems. The flowers are small, light yellow, with many stamens and pistils in the center, each pistil producing a single seed. The plant grows at 3,000 to 6,000 feet, along the western edge of the Colorado Desert and in the mountains of the Mojave Desert as far north as Inyo County. It flowers from April to August.

FIGURE 126. POTENTILLA

Another member of the family is the shrub ANTELOPE BUSH (*Purshia glandulosa*), figure 127, of dry slopes mostly between 3,000 and 9,000 feet, from Mono County to Arizona and Lower California. The twigs and leaves are glandular, the latter slightly woolly beneath. Petals are cream to yellowish, about one-fourth inch long. The one-seeded fruit is tipped with the short style and blooms from April to June. See also pages 34, 68.

FIGURE 127. ANTELOPE BUSH

CATCLAW (*Acacia Greggii*), figure 128, is our only native California member of the immense genus *Acacia*, which has almost universal occurrence in dry warm parts of the world. It has a pod typical of the Pea Family, but the flowers are small, in cylindrical spikes, and each with similar petals and many stamens. It is a straggling deciduous shrub with short recurved spines and leaves with four to six pairs of leaflets. It is found in washes and canyons below 6,000 feet in the Mojave and Colorado deserts and far beyond. It blooms from April to June. Anyone who has gotten entangled in Catclaw branches will agree that this name is most fitting.

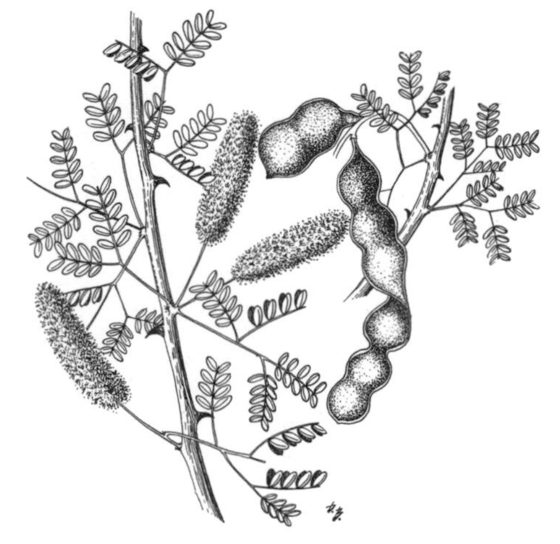

FIGURE 128. CATCLAW

Two other members of the Pea Family with flowers consisting largely of a tuft of stamens are in the genus *Prosopis*. One is SCREW-BEAN MESQUITE OR TORNILLA (*Prosopis pubescens*), figure 129, a shrub to a

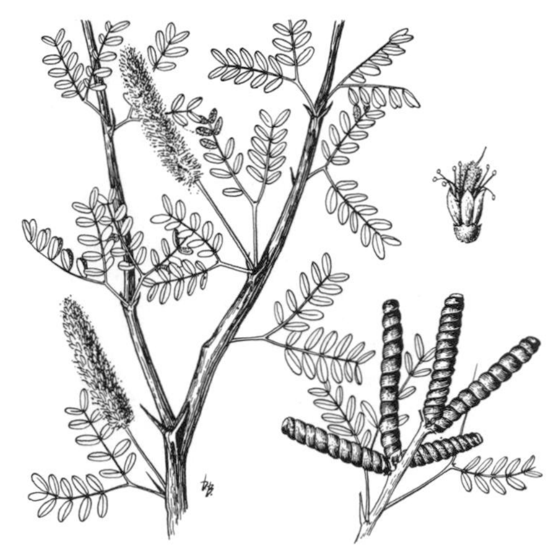

FIGURE 129. SCREW-BEAN MESQUITE

small tree. The foliage is somewhat grayish, the number of leaflets five to eight pairs. The small yellowish flowers are not pealike and are in spikes two to three inches long. The outstanding feature is the persistent coiled pod which is wound into a tight springlike cylinder an inch or more long. Found in canyons and washes below 2,500 feet, from the Colorado and Mojave deserts to western Fresno County and to Texas and Chihuahua, Screw-Bean blooms from May to July. See page 35.

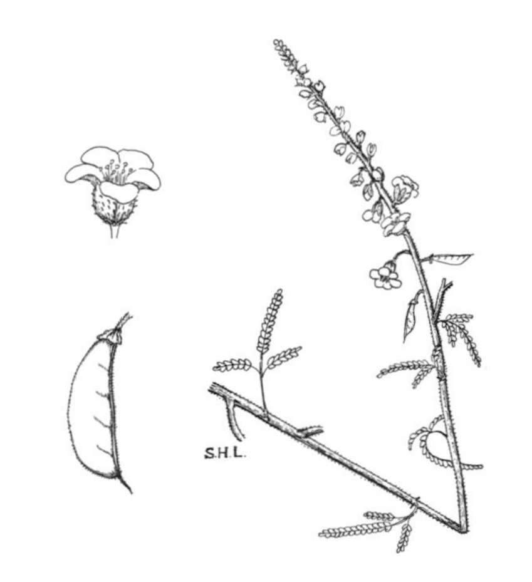

FIGURE 130. HOFFMANNSEGGIA

Another yellow-flowered member of the Pea Family, but with a more highly developed corolla (still not typically pealike, however) is *Hoffmannseggia microphylla*, figure 130, a subshrub two to three feet high. Conspicuous and not rare, it must have a common name, but I do not know one. It has many rushlike stems with small scattered leaves divided into numerous tiny leaflets. The stems end in long racemes of yellow to orange-red flowers, each of which may produce a flat pod up to one inch long. This plant is common about canyons and washes below 4,000 feet, in the Colorado Desert and then ranges to Lower California and Sonora. It blooms from March to May.

FIGURE 131. DESERT DEERWEED

DESERT DEERWEED OR ROCK-PEA (*Lotus rigidus*), figure 131, is a member of the Pea Family with truly pealike flowers, the upper petal forming an erect "banner," the two lateral petals "wings," and the two lower a "keel," in which are contained the stamens and pistil. This species is perennial, erect, woody at the base, one to three feet tall and bears at the summit flowers one-half inch or more long. These are yellow with some red. It is common on dry slopes and in washes, below 5,000 feet, from Inyo County to Lower California, Arizona, and Utah. It blooms from March to May.

The most conspicuous shrub on the desert is CREOSOTE BUSH or GREASEWOOD *(Larrea divaricata)*, figure 132, with its waxy olive-green leaves, each of which consists of a pair of leaflets, and with its yellow flowers having five petals partly twisted like the vanes of a windmill. The fruit is globose, white-hairy, and separates at maturity into five one-seeded parts. An open shrub, from three to twelve feet tall, Creosote Bush is the dominant plant over vast areas, ascending to about 5,000 feet and ranging from coastal slopes in western Riverside County through the deserts to Utah, Texas, and Mexico. Flowering is in April and May. It is quite resinous and gives off a penetrating odor especially after rain. See page 37.

FIGURE 132. CREOSOTE BUSH

In the same family (Caltrop) belongs the introduced weed so commonly called Puncture-Vine because of the heavy spines on its five-parted fruit. A close relative is a desert native, KALLSTROEMIA *(K. californica)*, figure 133, a prostrate annual forming mats which may actually carpet the desert for miles after a rainy summer. The yellow flowers are small; the fruit does not have spines but merely tubercles. It and a closely related species (*K. parviflora*) bloom from August to October and are found mostly from the eastern parts of the California deserts to Mississippi and Mexico.

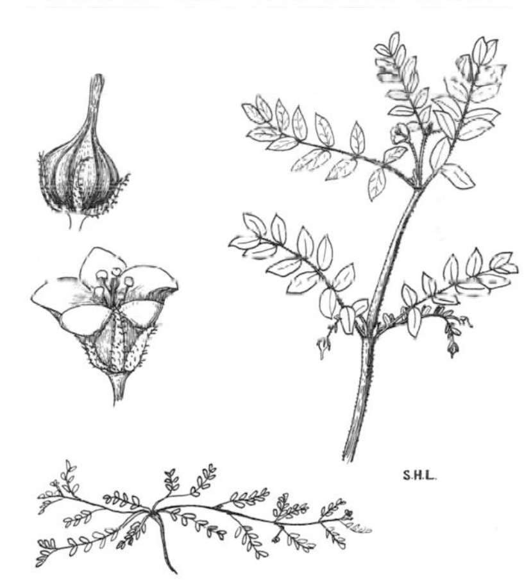

FIGURE 133. KALLSTROEMIA

CRUCIFIXION THORN *(Holacantha Emoryi)*, figure 134, is a rigid, much-branched thorny shrub with small leaves that are soon shed. The flowers are borne singly or in clusters on the heavy thornlike branches and have seven or eight sepals and as many petals. The pistils are five to ten and almost separate. The shrub is three to eight feet tall and is occasional in gravelly places and on dry plains across the southern Mojave Desert and the

FIGURE 134. CRUCIFIXION THORN

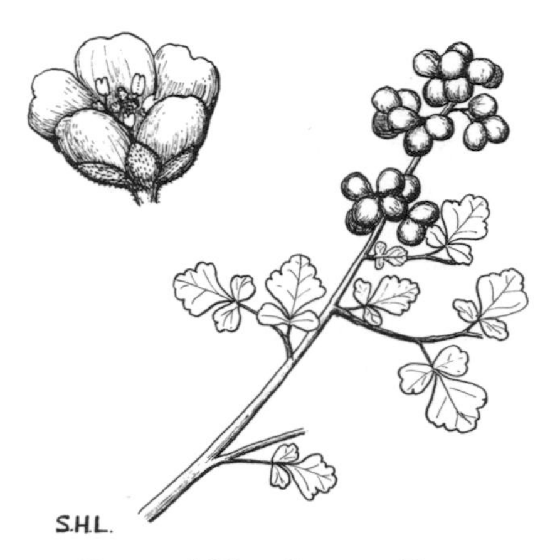

FIGURE 135. SQUAW BUSH

FIGURE 136. FELT PLANT

FIGURE 137. BLAZING STAR

northern Colorado Desert to Arizona. It is in flower from June to July.

The Sumac Family is largely one of warm regions and is characterized by resinous or milky sap, often poisonous, as in the case of Poison-Ivy and Poison-Oak. SQUAW BUSH *(Rhus trilobata),* figure 135, is in the same genus with Sumac and Lemonadeberry, which latter it resembles in having reddish fruits with an acid covering. The form on the desert is mostly a low shrub with three main leaflets to each leaf and with rather a strong odor when crushed. The small yellowish flowers are in clustered spikes. It occurs in dry often rocky places between 3,500 and 5,500 feet, from the mountains of the Mojave and northern Colorado deserts to Utah and Arizona, blooming from March to April.

Related to the Hollyhock is *Horsfordia Newberryi,* figure 136, for which I have seen the name FELT PLANT. It is rather woody, three to seven feet tall, with heart-shaped felty-woolly leaves one to three inches long. The petals are yellow, rounded, one-third of an inch long. The row of carpels (divisions of the fruit) has wings projecting above the calyx. This plant is found in dry rocky places below 2,500 feet, in the western Colorado Desert and to Lower California and Sonora. It flowers from November to December and March to April.

The BLAZING STAR of the same family as Sandpaper Plant (see pages 38, 72) has similar barbed hairs and white to yellow flowers. The one shown in figure 137, (*Mentzelia nitens*) is an annual with white shining stems, mostly lobed leaves, and bright yellow petals from one-half to almost one inch long. The ovary is borne below the flower and is elongate. Stamens are many. It inhabits sandy and gravelly places below 5,000 feet, from Mono

County to Riverside County and to Arizona and Utah. It is a spring bloomer.

California deserts have a good many cacti, among them NIGGERHEADS *(Echinocactus polycephalus),* figure 138, with several subglobose or subcylindric stems a foot or so thick and in clumps of ten to thirty. Each stem usually has about eighteen to twenty ribs, each with numerous spine-clusters. The yellow flowers are one to two inches long; the fruit is densely woolly. This cactus is found in well-drained places between 2,000 and 5,000 feet, in the northern Colorado Desert and much of the Mojave Desert and blooms from March to May. Another Echinocactus is *E. acanthodes,* the common Barrel Cactus with taller mostly solitary stems and the fruits not woolly.

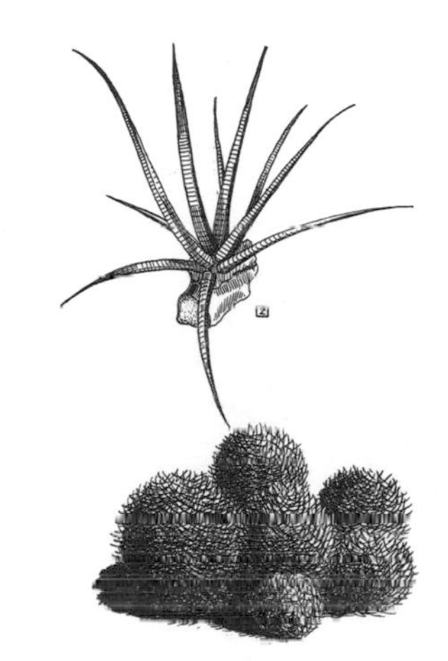

FIGURE 138. NIGGERHEADS OR COTTONTOP CACTUS

A common Evening-Primrose on the dry, and often in the rocky, places of the desert is *Oenothera primiveris,* figure 139. It is for the most part a stemless winter annual with a long taproot, a rosette of leaves, and yellow flowers one to three inches high. The petals are yellow, opening toward evening, aging orange-red as the next day advances. It is found mostly below 5,000 feet and extends from the northern and eastern parts of the Colorado Desert to Inyo and Kern counties, then to Utah and Texas. Flowers are from March to May. See pages 42, 43.

FIGURE 139. EVENING-PRIMROSE

Another Evening-Primrose is *Oenothera brevipes,* figure 140, an annual with mostly basal leaves just above which the stem is spreading-hairy. It ends above in a nodding tip with bright yellow flowers about one inch across. The capsule is elongate and spreading. This common desert annual is found in sandy washes and on dry flats below 5,000 feet, from Inyo and western San Bernardino counties south and east to Imperial County

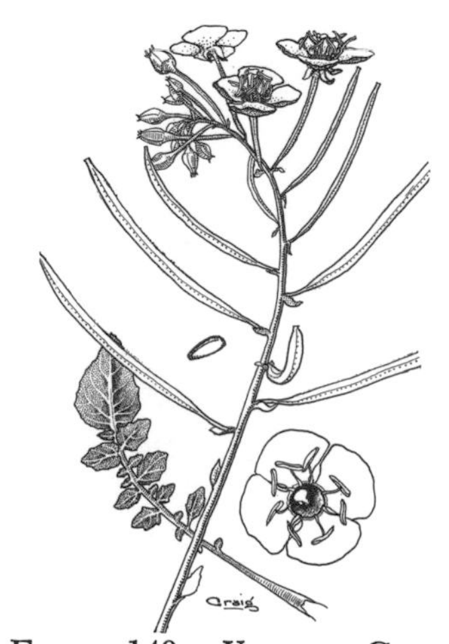

FIGURE 140. YELLOW CUPS

FIGURE 141. CYMOPTERUS

and to Arizona and Nevada. A handsome plant, blooming from March to May, it is sometimes called YELLOW CUPS.

In the Carrot Family (see page 20), one of the most easily recognized groups by virtue of the longitudinally winged fruits is CYMOPTERUS *(C. panamintensis)*, figure 141, a stemless heavy-rooted perennial forming large tufts of finely divided leaves and umbels (flower clusters with all the branches radiating from the same point) of small greenish-yellow flowers. The fruits are about one-third inch long. Like most members of the family, it is aromatic when crushed. It grows in dry rocky places between 2,000 and 8,000 feet, in the mountains of the Mojave Desert, and flowers from March to May.

FIGURE 142. CLIMBING MILKWEED

A CLIMBING MILKWEED *(Sarcostemma hirtellum)*, figure 142, is like other milkweeds (see page 74) in its milky sap and paired leaves, but is a twining slender-stemmed plant. The flowers are greenish-yellow, about one-sixth of an inch long, and the slender pods an inch and one-half or more. It is found, often climbing over other plants, in washes below 3,500 feet, in the Colorado and eastern Mojave deserts and in adjacent Nevada and Arizona. Flowers are in bloom from March to May. A related species less hairy and with purplish flowers *(S. cynanchoides)* ranges from the desert to the coast.

FIGURE 143. GOLDEN GILIA

In the Phlox Family (see pages 21, 44, 84) Linanthus is characterized by paired leaves divided into linear segments. A yellow-flowered species, GOLDEN GILIA *(Linanthus aureus)*, figure 143, is a low plant, usually several-branched, with leaves rather remote. The flowers have an orange to brownish-purple throat and are one-fourth to one-half inch long. It is locally common over large areas in sandy places below 6,000 feet, in both deserts and occasionally in the coastal valleys, blooming from March to June. A whitish form occurs with the yellow one.

In the Borage Family (see pages 75 and 86) is the very characteristic FIDDLENECK (*Amsinckia tessellata*), figure 144, with yellow to orange flowers, stiff prickly hairs, and coiling flower clusters. Color of flowers and their size vary with the different species, but this most characteristic desert one tends to be orange and less than half an inch long. The back of the nutlet or "seed" is tessellate like a mosaic. The species is common in dry mostly sandy and gravelly places below 6,000 feet, throughout the deserts and along the Inner Coast Ranges to Contra Costa County; east of the Sierra Nevada it ranges to Washington. Flowers appear between March and June.

FIGURE 144. FIDDLENECK

A SNAPDRAGON (*Antirrhinum filipes*), figure 145, is a small annual with threadlike stems, small leaves, and yellow flowers about half an inch long. It twines among low bushes in sandy places below 5,000 feet, from Inyo County south through both deserts and east to Utah and Nevada. Flowers may be found from February to May depending on the elevation. Both main stems and pedicels (the small stems bearing individual flowers) are capable of twining. It is in the same family as Paint-Brush; see pages 22, 51.

FIGURE 145. SNAPDRAGON

CALABAZILLA OR WILD GOURD (*Cucurbita foetidissima*), figure 146, is an amazing plant. It is a large strong-smelling perennial from an immense fusiform root and has long trailing stems bearing erect leaves to almost a foot high. The flowers are often about four inches long, the female producing striped gourdlike fruits three to four inches in diameter. The range extends from the San Joaquin Valley to San Diego and across the Mojave

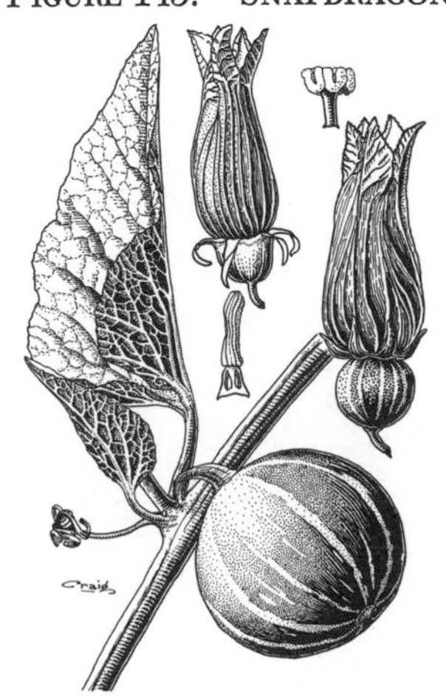

FIGURE 146. WILD GOURD

FIGURE 147. BEDSTRAW

FIGURE 148. BRICKELLIA

FIGURE 149. GOLDENHEAD

Desert to Nebraska and Texas. It flowers from June to August. Two smaller-leaved and smaller-flowered species also occur.

Bedstraw is a name applied to a group of slender-stemmed plants with whorls of small leaves and minute usually four-parted flowers. One of the most common desert species of BEDSTRAW *(Galium stellatum)*, figure 147, is bushy, much-branched above the woody base, one to two feet high, rough-pubescent, and with small narrow leaves. The male flowers are in crowded clusters, the female at the ends of little branchlets. The fruit is soft-hairy and small. Common on dry rocky slopes below 5,000 feet, the plant ranges from our deserts to Arizona and Nevada and blooms in March and April.

In the Sunflower Family (see pages 52, 77) there are many yellow-flowered plants, especially when all the florets in a head are the tubular kind such as we find in the center of the Sunflower itself. Such is the case in BRICKELLIA *(Brickellia arguta)* figure 148, a much-branched shrub a foot or more tall, with zigzag stems and aromatic, bright green, toothed leaves. The heads are about one-half inch high with a number of small yellow flowers. It grows in rocky places below 4,500 feet, from Inyo County to northern Lower California, blooming in April and May.

GOLDENHEAD (*Acamptopappus sphaerocephalus*), figure 149, is another member of the Sunflower Family with numerous small yellow florets in each head. The plant is a low round-topped shrub, much-branched, with narrow leaves about one-half inch long and somewhat shorter rounded heads. Flowering is from April to June. Goldenhead occurs on the open desert below 4,000 feet, from Lone Pine, Inyo County, to eastern San Diego County and across the desert to Utah and Arizona.

Another closely related species (*A. Shockleyi*), color plate 76, has petallike ray-flowers at the edge of the head.

GOLDENBUSH (*Haplopappus linearifolius*), figure 150, is closely related to Goldenhead and has yellow ray-flowers as well as the many tubular central ones. It is a much-branched more or less flat-topped shrub, two to four feet tall, resinous, narrow-leaved, with heads about one inch in diameter. It is common on dry slopes and banks below 6,000 feet, from Butte and Lake counties south along the lower mountains and crossing the deserts to Utah and Arizona. Flowering from March to May, its showy heads add much to the interior hills of southern California. See page 52.

FIGURE 150. GOLDENBUSH

RABBITBRUSH consists of a series of low shrubs of dry plains and slopes, often in subalkaline places of interior California and the states about the Great Basin, ranging north to British Columbia and east to South Dakota. The rayless narrow heads borne in numerous clusters, the usually narrow leaves, the stems often invested in very tight wool, the rather resinous odor of the foliage, and the fragrance of the flowers—all these traits are characteristic. Furthermore, it blooms in late summer and fall. Rabbitbrush (*Chrysothamnus nauseosus*), figure 151, occurs in many different forms and ascends the mountains to about 9,000 feet.

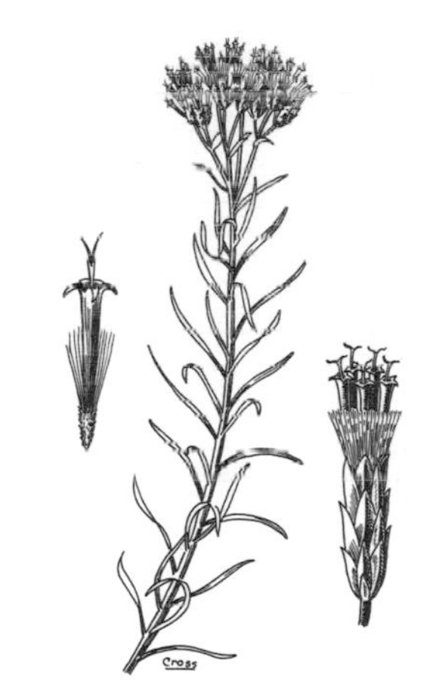

FIGURE 151. RABBITBRUSH

ROCK-GOLDENROD (*Petradoria pumila*), figure 152, is sometimes included in Goldenrod itself (*Solidago*). It is a low stiff plant, resinous, light green, with leafy stems and with narrow leaves two or three inches long, the leaves having three to five longitudinal veins. The heads are few-flowered, with one to three short rays as well as the central tubular flowers. It is found on dry limestone hillsides, at

FIGURE 152. ROCK-GOLDENROD

FIGURE 153. SWEETBUSH

FIGURE 154. SUNFLOWER

FIGURE 155. DESERT-SUNFLOWER

3,500 to 7,000 feet, in the mountains of the eastern Mojave Desert and ranges to Wyoming and Texas. The flowers appear from July to October.

SWEETBUSH *(Bebbia juncea)*, figure 153, is a diffuse shrub two to three feet high, leafless or with narrow quickly deciduous leaves. The heads have tubular flowers only and are very sweet-scented. It is found on gravelly fans, rocky washes, and canyon sides below 4,000 feet, from the White Mountains to Lower California and east to New Mexico, but is most common in the Colorado Desert. It occurs also in interior valleys of coastal drainage; flowering is from April to July.

The SUNFLOWER is known to all of us, but many of us do not realize that there are many different species, hence it is not too surprising to find the desert with its quota. One of these is *Helianthus petiolaris* var. *canescens*, figure 154, with a grayish foliage because of the stiff white hairs. Normally rather low, it can be as much as three or more feet tall, with leaves two or three inches long. The flower heads are one to two inches across and appear from March to June. Found in the open desert, especially in sandy places like Borrego Valley and east of Imperial Valley, it ranges to Texas and Mexico.

DESERT-SUNFLOWER (*Geraea canescens*), figure 155, is the common name of an attractive annual one to two feet high, related to the true Sunflower. It is glandular, has white hairs and toothed leaves. The heads are solitary or in panicles, almost two inches across, with ten to twenty golden rays around a yellow disk. It is common (usually with Sand-Verbena and Desert-Primrose) on sandy flats below 3,000 feet, in the Colorado and eastern Mojave deserts and to Utah and Sonora. It blooms mostly from February to May and more sparingly in October and November.

A close relative of the Sunflower is Goldeneye *(Viguiera deltoidea* var. *Parishii),* figure 156, a rounded subshrub one to three feet high, many branched, with harsh stems and more or less triangular leaves. The heads are borne on long naked stems and are over an inch in diameter, with yellow ray-flowers. It is found in sandy desert canyons and on mesas below 5,000 feet, in the Colorado Desert and eastern Mojave Desert, reaching the coast near San Diego and eastward into Nevada and Arizona. It flowers from February to June and again in September and October.

Brittle Bush or Incienso *(Encelia farinosa),* figure 157, is another relative of the Sunflower, with a woody trunk giving rise to many branches so as to form a compact low rounded bush one to two and one-half feet high and often several feet across. The leaves are silvery-gray, the heads yellow or brown in the center, and the rays yellow. It is common about washes and on stony slopes below 3,000 feet, from Kern River Canyon and other hot interior valleys to San Diego County and over most of the California deserts to Utah and Sinaloa. It blooms from March to May. The deserts have other species with greener leaves.

Coreopsis *(C. Bigelovii),* figure 158, is common as several different species in our California flora. This one is an annual with naked stems arising from a mass of lobed leaves and varies from a few inches to almost two feet in height. It grows on dry gravelly hillsides between 1,000 and 5,000 feet, from southern Monterey and Tulare counties south and across much of the western Mojave Desert. It blooms from March to May. Some

Figure 156. Goldeneye

Figure 157. Incienso

Figure 158. Coreopsis

FIGURE 159. BURROWEED

FIGURE 160. BUR BUSH

FIGURE 161. PAPERFLOWER

other annual species differ from it only in technical features; see page 54.

One group in the Sunflower Family has inconspicuous heads, the male or staminate consisting of small tubular flowers and the female or pistillate possessing one to four pistils surrounded by a burlike outer involucre. Such a plant is BUR-SAGE or BURROWEED *(Franseria dumosa)*, figure 159, a low grayish subshrub one to two feet high having divided leaves. The stem tips bear the heads of flowers and the burlike female heads with sharp spines often persist for a long time. It is one of the most common plants of the desert, often growing in between plants of Creosote Bush (page 97) at below 3,500 feet, and ranges to Utah and Sonora. It blooms from February to June and September to November.

Another BUR BUSH *(Franseria eriocentra)*, figure 160, is larger both in stature and leaf. The leaves are gray-woolly, at least when young; the drawing shows the male heads above and the woolly female ones below in the inflorescence. It grows on dry slopes at between 2,500 and 5,000 feet, in the eastern Mojave Desert and to Utah and Arizona, flowering from March to May.

PAPERFLOWER *(Psilostrophe Cooperi)*, figure 161, is well named since the yellow ray-flowers persist and become dry and papery in age. The plant is somewhat woody at the base, many stemmed, white-woolly (especially on the younger parts), and has simple undivided leaves one to almost three inches long. Inhabiting rocky desert mesas and sandy fans between 2,000 and 5,000 feet, it is found in the eastern Mojave Desert from the Kingston and Clark mountains to the Little San Bernardino Mountains and in the northern Colorado Desert. It responds to both spring and summer rains and in the proper year blooms twice, once in spring and once in autumn. See page 55.

Rock-Daisy *(Perityle Emoryi)*, figure 162, is a low glandular annual, with brittle branches and broad, toothed to lobed leaves. The heads are small, with a yellow center and whitish to yellowish rays which are not very conspicuous. The plant is common in rock-crevices and among boulders below 3,000 feet, through most of the desert and extending its range along the immediate coast as far north as Ventura County. Eastward it occurs to Nevada and Sinaloa. Flowers appear as early as February and as late as June.

Yellowhead *(Trichoptilium incisum)*, figure 163, is another annual that is loosely white-woolly and aromatic. It is low, diffusely branched, with sharply toothed leaves. The yellow heads have many small florets, all tubular. This species is common on desert pavement or in sandy places below 2,200 feet, in the Colorado and southern Mojave deserts and ranges into southern Nevada, western Arizona, and northern Lower California. Flowers can be found from February to May and October to November, if the rains come at the right times.

Desert-Marigold occurs as three different species in the California deserts. One *(Baileya pauciradiata)* has small pale yellow heads, each with only five to seven ray-flowers. A second *(Baileya multiradiata)*, page 56, has twenty to fifty ray-flowers and the leaves near the base of the plant. It is found only in the eastern Mojave Desert. The third (*Baileya pleniradiata)*, figure 164, also has numerous rays, but with the leaves extending farther up on the stems. It is widespread in sandy places below 5,000 feet, the gray foliage and yellow heads making an attractive

Figure 162. Rock-Daisy

Figure 163. Yellowhead

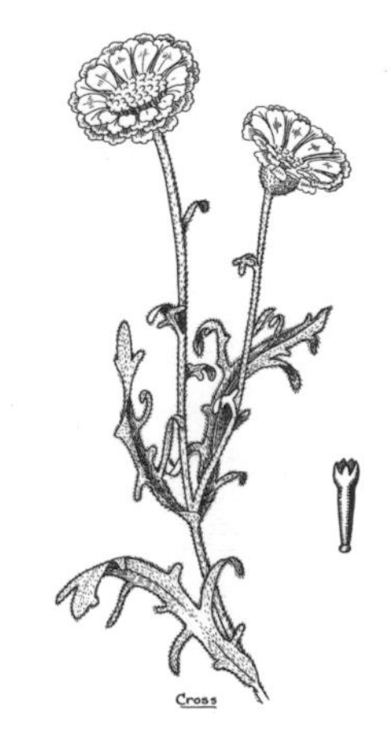

Figure 164. Desert-Marigold

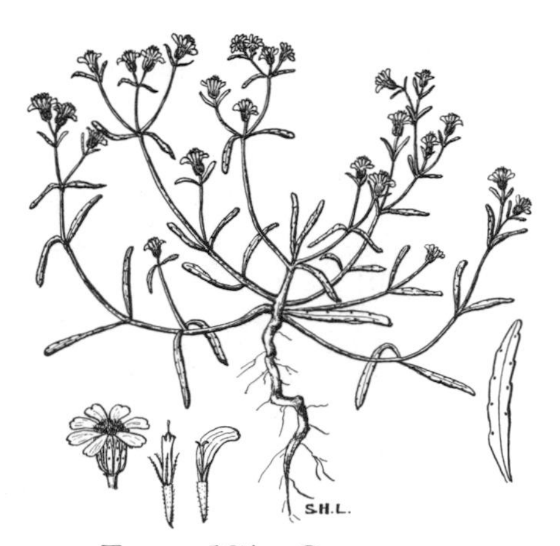

Figure 165. Chinchweed

Figure 166. Dyssodia

Figure 167. Groundsel

plant. Flowers appear from March to May and October to November.

After summer rains the desert floor, especially on clayey and sandy flats below 5,000 feet, may be carpeted with a smelly little annual called Chinch-weed *(Pectis papposa)*, figure 165. The narrow leaves are opposite and the small heads of florets are yellow. The plants are dotted with oil-glands. It ranges from the Colorado Desert to the Death Valley region, then eastward to Utah, New Mexico and northern Mexico. It was once used by the desert Indians as flavoring in their cooking and for its odor.

Another plant with conspicuous oil-glands, especially on the involucre surrounding the flower-heads, is a strong-scented perennial with a woody base and thick narrow leaves less than an inch long. The heads have orange-yellow erect rays that may turn purplish; sometimes these are quite lacking and only the central tubular florets are present. Dyssodia *(D. porophylloides)*, figure 166, grows in sandy washes and on mesas and rocky slopes at below 3,500 feet, from the western borders of the Colorado Desert to the southern Mojave Desert and into Arizona and Sonora.

Groundsel *(Senecio stygius)*, figure 167, is a very pretty little plant of a large group, the genus *Senecio* consisting perhaps of a thousand species. It is characterized by its cylindrical involucre around each head of elongate equal bracts, with a row of much shorter basal ones. The species is greenish in foliage, perennial, up to about one foot high, with bright yellow heads about one inch in diameter. It is frequent on dry slopes between 4,000 and 6,500 feet, in the mountains of the eastern Mojave Desert and into Nevada and Arizona. It blooms from

April to May. For a colored illustration see page 56.

Another Groundsel *(Senecio Douglasii* var. *monoensis)*, figure 168, is more or less woody at the base, bushy, one to three feet tall, yellowish-green, with flat linear leaves or these sometimes dissected into linear segments. The heads are about one inch across and clear yellow. Found on dry slopes and in washes between 2,000 and 6,500 feet, it occurs in the northern Colorado Desert and in the Mojave Desert north to Mono County. It is found as far east as Utah and Arizona, blooming from March to May and sometimes also in the fall.

Pigmy-Cedar or Desert-Fir *(Peucephyllum Schottii)*, figure 169, is an aromatic shrub up to eight feet high, with terminal tufts of bright green crowded leaves up to one inch long. The heads have no rays, merely tubular central flowers which are quite yellow, although their tips may turn purplish in age. It grows in rocky places and canyons below 3,000 feet, in the Colorado and eastern Mojave deserts, also in Nevada, Arizona, and northern Lower California. Flowering is from December to May.

Cotton Thorn *(Tetradymia comosa)*, figure 170, is a rigidly branched bush two to four feet tall. It is densely white-woolly, with linear leaves that may become modified into long slender sharp structures. The species illustrated here has the heads in small clusters; a more common one on the deserts (*Tetradymia axillaris*), page 57, has them more remote from each other and on separate stems. In either case the heads are bright yellow. Common on dry slopes below 6,400 feet, these two species can be found on the Mojave Desert as far north as Mono County and in the mountains bordering

Figure 168. Groundsel

Figure 169. Pigmy-Cedar

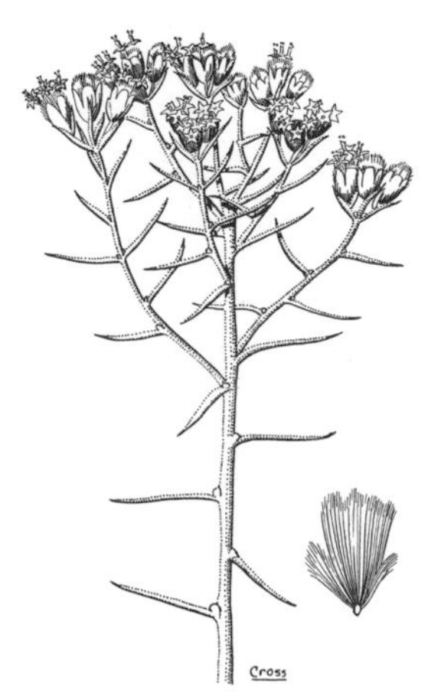

Figure 170. Cotton Thorn

FIGURE 171. TRIXIS

FIGURE 172. GLYPTOPLEURA

the western Colorado Desert. Other species greatly resemble these.

TRIXIS *(T. californica)*, figure 171, is in the Sunflower Family also, but in a group much more common in the Southern Hemisphere than in the Northern, having each floret in the head two-lipped. It is a shrub about three feet high, leafy to the heads, and quite glandular. The leaves are one to two inches long, the heads about one-half inch long. It is frequent in canyons and washes below 3,000 feet and occurs in the Colorado Desert and north in the Mojave Desert as far as the Sheephole Mountains. It ranges eastward to western Texas and northern Mexico. Flowers can usually be found from February to April.

GLYPTOPLEURA is a dandelionlike plant, low, almost stemless, and with all the florets in the head raylike or strap-shaped. Our common species, *G. setulosa*, figure 172, has curly lobed leaves with a whitish margin and many teeth. The flower heads are about one and one-half inches across and creamy yellow. Glyptopleura is a plant of sandy flats at 2,000 to 3,500 feet, from the western Mojave Desert to southern Utah and northwestern Arizona. It is really a charming little thing, blooming in April and May.

Topographic Map of California

COUNTY MAP OF CALIFORNIA

INDEX TO COLOR PLATES

(References are to plate numbers)

FLOWERS REDDISH

FLOWERS WHITE TO GREENISH

Flowers Blue to Lavender

Flowers Yellow to Orange or Cream

INDEX

(References are to page numbers)